Colonialism, Culture, Whales

Environmental Cultures Series

Series Editors:
Greg Garrard, University of British Columbia, Canada
Richard Kerridge, Bath Spa University

Bloomsbury's *Environmental Cultures* series makes available to students and scholars at all levels the latest cutting-edge research on the diverse ways in which culture has responded to the age of environmental crisis. Publishing ambitious and innovative literary ecocriticism that crosses disciplines, national boundaries, and media, books in the series explore and test the challenges of ecocriticism to conventional forms of cultural study.

Titles available:
Bodies of Water, Astrida Neimanis
Cities and Wetlands, Rod Giblett
Civil Rights and the Environment in African-American Literature, 1895–1941, John Claborn
Climate Crisis and the 21st-Century British Novel, Astrid Bracke
Colonialism, Culture, Whales, Graham Huggan
Ecocriticism and Italy, Serenella Iovino
Fuel, Heidi C. M. Scott
Literature as Cultural Ecology, Hubert Zapf
Nerd Ecology, Anthony Lioi
The New Nature Writing, Jos Smith
The New Poetics of Climate Change, Matthew Griffiths
This Contentious Storm, Jennifer Mae Hamilton

Forthcoming titles:
Anthropocene Romanticism, Kate Rigby
Cognitive Ecopoetics, Sharon Lattig
Eco-Digital Art, Lisa FitzGerald
Teaching Environmental Writing, Isabel Galleymore

Colonialism, Culture, Whales

The Cetacean Quartet

Graham Huggan

BLOOMSBURY ACADEMIC
LONDON • NEW YORK • OXFORD • NEW DELHI • SYDNEY

BLOOMSBURY ACADEMIC
Bloomsbury Publishing Plc
50 Bedford Square, London, WC1B 3DP, UK
1385 Broadway, New York, NY 10018, USA

BLOOMSBURY, BLOOMSBURY ACADEMIC and the Diana logo are trademarks of Bloomsbury Publishing Plc

First published in Great Britain 2018

Cover design: Burge Agency
Cover image © Shutterstock

A catalogue record for this book is available from the British Library.

Library of Congress Cataloging-in-Publication Data
Names: Huggan, Graham, 1958, author.
Title: Colonialism, culture, whales : The cetacean quartet / Graham Huggan.
Description: London, UK : Bloomsbury Publishing Plc, 2018. |
Series: Environmental Cultures series | Includes bibliographical references and index.
Identifiers: LCCN 2018002184 (print) | LCCN 2018006674 (ebook) |
ISBN 9781350010901 (ePUB) | ISBN 9781350010918 (ePDF) |
ISBN 9781350010895 (hardback)
Subjects: LCSH: Whales. | Whaling–History. | Whale watching. | Whales–Conservation.
Classification: LCC QL737.C4 (ebook) | LCC QL737.C4 H84 2018 (print) | DDC 599.5–dc23
LC record available at https://lccn.loc.gov/2018002184

ISBN: HB: 978-1-3500-1089-5
ePDF: 978-1-3500-1091-8
eBook: 978-1-3500-1090-1

Series: Environmental Cultures

Typeset by Integra Software Services Pvt. Ltd.

To find out more about our authors and books visit www.bloomsbury.com and sign up for our newsletters.

Contents

Acknowledgements

Versions of Chapters 1 and 2 have previously appeared in the following, slightly modified forms: 'Last Whales: Eschatology, Extinction, and the Cetacean Imaginary in Winton and Pash', *Journal of Commonwealth Literature* 52.2 (2017): 382–396; 'Killers: Orcas and Their Followers', *Public Culture* 29.2 (2017): 287–309. Thanks to the publishers for permission to reprint. The author and publisher would also like to thank the following for permission to use the inside images in this book: 'Cheynes whale' by Paul Fearn (p.1) © Paul Fearn/Alamy Stock Photo; 'William Scoresby Arctic Navigator. Date: 1760–1829 (p.25, left) © Chronicle/Alamy Stock Photo; 'William Scoresby Jr' by Edwin Cockburn (p.25, right) © Bridgeman Images; 'Tilikum and his trainers at SeaWorld in 2011' by Phelan Ebenhack (p.57) © Shutterstock; 'Scar, the morning of February 9, 2009' by Bryant Austin (p.85) © Bryant Austin.

All books, including monographs, are collaborative enterprises. This one would not have been possible without the academic support of the HERA 'Arctic Encounters' team, especially Roger Norum, and, more recently, that of the Environmental Humanities doctoral training network (ITN), whose excellent students should be an inspiration to us all. Many thanks also to Francis O'Gorman, who patiently read and commented on all of the chapters, and to Greg Garrard and Richard Kerridge for believing in the project in the first place. Finally, life partners are just that – with you through thick and thin – so the last word must go to Bine, with my love.

Preface

Most stories of cruelty have heroic and melancholic versions; and so it is with the history of human encounters with whales. This book opts largely for the melancholic view as a basis for its scattered reflections on whales in the overlapping if non-identical contexts of imperialism, industrial capitalism, and post-industrial corporate power. Whales, after all, are a symbol of guilt as well as an opportunity for atonement, and while they register concern for the future, they also offer painful reminders of a violent past.

One of the premises of the book is that this past is tied up in, though not necessarily reducible to, empire. In the four loosely connected chapters that follow, I will be observing Stephen Howe's basic distinction between *empires* as large-scale political units; *imperialism* as the actions and attitudes that justify the creation and consolidation of those units; and *colonialism* as more specific 'systems of rule by one group over another, where the first claims the right (a "right" usually established by conquest) to exercise exclusive sovereignty over the second and to shape its destiny' (2002: 30–31). That whaling was a service industry for empires – by no means just the British one – is well documented. As Philip Hoare observes, in many ways whales laid the foundations for the British Empire as well as providing the commercial basis for some of the first colonial industries (Hoare 2008). Expanding on this, Richard Ellis notes the close association of whaling with European journeys of discovery and exploration; the formative role of whaling in the early establishment and development of colonies; and the crucial importance of the whaling industry in exporting European and, later, American culture and ideas to the wider world (Ellis 1999; see also Crosby 2009).

This is not to reduce whaling to a mere instrument of empire, in either its economic or cultural forms, but to suggest that it often provided a persuasive rationale for the building of empires (Dodds 2002). As might be deduced from this, it was also at the heart of imperial rivalries played out between the major European maritime powers. More recently, as will be seen, whales have played their part as well in what the Norwegian anthropologist Arne Kalland calls the 'symbol-politics' that has come to surround the global environmental movement – a movement in which cetacean conservation has turned into a high-stakes political gambit, a competitive exercise in international relations and corporate lobbying power (Kalland 2009: 193; see also DeSombre 2007, Epstein 2005). This scenario is sometimes seen in terms of the 'colonial-style discrimination' (Sciullo 2008: n.p.) with which powerful Western states have forced whale-hunting bans onto formerly colonized countries, or have lodged bitter protests against aboriginal subsistence whaling, with this last – a point to which I will return – emphasizing the key role of culture, or perhaps more accurately *perceptions* of culture, within both whaling and anti-whaling debates.

Culture has played a significant role as well in shaping human attitudes towards cetaceans: for, down through the centuries, people the world over have been much more likely to encounter cetacean representations than actual living whales (Kalland 2009). These representations have often organized themselves around specific *narratives*. As subsequent chapters will show, narratives about whales take many different forms, many of them either explicitly or implicitly allegorical; for whales, whose existence long predates ours, have frequently been associated with mythical stories of human origin as well as apocalyptic presentiments of planetary demise. By no means all of these narratives are sad, but several of them have melancholic tendencies. The story of commercial whaling, in particular, is now almost universally looked upon as one of 'unrelieved greed and insensitivity. In no other activity has our species practiced such a relentless pursuit of wild animals, and if no whale species has become extinct at the hands of the whalers, it is not for the want of trying' (Ellis 1999: x). Whales have literally been torn apart to create oil for lighting, soap, and margarine; baleen and bone for various decorative and sartorial purposes. For a time, the trade in whales for oil would match the trade in humans for sugar as the commercial basis for the British Empire (Hoare 2009: 277); or as a character in Ian McGuire's 2016 novel *The North Water* cynically puts it, 'And without the Empire who would buy the whalebone, who would buy the oil?' (McGuire 2016: 14).

Many representations, 'renderings' of a different kind (Shukin 2009), have been almost as brutal as the whaling industry itself, with whales being cast as demonic presences, malevolent portents of the approaching end times. One of the many ironies surrounding whales is that they helped further the same Industrial Revolution that hastened their destruction (Hoare 2008: n.p.). In this and other ways, they are both symbols and symptoms of a Western-style industrial modernity – for which colonialism is the flip side – that has created the conditions for its own unravelling (Mignolo 2000).[1] The tense that best captures these conditions is the future anterior. Scottish nature writer Kathleen Jamie, wandering through the Whale Hall in Bergen's Natural History Museum, marvels at the decaying bones of 'twenty-four cetacean skeletons crowded under the ceiling' (2012: 96). One whale skeleton alone, that of a gigantic blue, is 'less an animal, more a narrative' (97). The different cetacean narratives add up to a devastating commentary to which even words such as 'waste' and 'slaughter' and 'holocaust' and 'shame' cannot do full justice (111). Jamie duly joins a small international team of conservators who lovingly polish up the bones, dedicated to preserving a future past (112).

Like most books, this one has not turned out the way I intended it. Originally my thought was to adopt a broadly postcolonial approach to the various histories surrounding whales, adding to the growing body of work on postcolonial ecology that offers 'more-than-human' readings of both the imperial present and the colonial past (see, for example, DeLoughrey and Handley 2011; DeLoughrey, Didur, and Carrigan 2015). To some extent, I also wanted to respond to Jonathan Steinwand's contention that there has been a 'cetacean turn in environmentalist iconography' (2011: 182) in which postcolonial literature has looked to whales and dolphins 'for guidance in how human animals participate in postcolonial ecology' (182) – an emphasis seen in such relatively recent works as Witi Ihimaera's *The Whale Rider* (1987), Zakes Mda's *The Whale Caller* (2005), Linda Hogan's *People of the Whale* (2008), and Amitav Ghosh's *The Hungry Tide* (2004). Four literary works alone hardly constitute a 'cetacean turn', but they do remind us, as Steinwand says, that postcolonial literature and criticism at its best offers powerful 'localized challenges to [an] environmentalist universalism' (184) that can be too sentimental for its own good, and that is given at times to illustrating the paradox that 'those who worry the most about the destruction of nature are usually those who make the problem worse' (193). Over and against this, Steinwand suggests, lies the trickster sensibility of cetaceans themselves, liminal creatures that demonstrate the qualities of what he calls (following Gerald Vizenor) 'indigenous survivance', using their 'uncanny otherness' (196) to remind indigenous and non-indigenous peoples alike of the need to 'move with the times and creatively adapt' (196; see also Vizenor 1999).

Steinwand's essay deals with represented whales, of course, but even so it struck me in reading it that a crucial *material* dimension of cetaceans was not being accounted for; that whales, in effect, were being used all over again as figures for desired human qualities (adaptability, creative transformation, etc.) and, as so often, were being turned into something other than themselves. This led me to look again at Donna Haraway's influential work, for example *When Species Meet* (2008), which has set a high bar for 'material-semiotic' approaches to animal studies, in which the material presence of animals is acknowledged even as their symbolic instrumentality is also recognized, their capacity to function across different human societies and cultures as multivalent signs. Haraway's work has been influential, too, in establishing a set of first principles for environmental humanities, a critical conglomeration of loosely fitting disciplines (literary studies, anthropology, geography, history, philosophy) designed to add cultural breadth and historical depth to wide-ranging environmental issues and problems that have tended by and large to be the domain of the social and natural sciences. Environmental humanities approaches are not intended to *replace* scientific ones of course, but rather to *complement* them, and a feature of work done so far in the field has been its practitioners' willingness to work across the disciplines while recognizing the limitations of their own 'core' disciplines. In this last sense, most of the best recent examples of environmental humanities criticism are collaborative work.

Perhaps inevitably, what began to emerge in *The Cetacean Quartet* was neither a single approach nor a committed line of argument, but rather a critical mix of celebratory eclecticism and material attentiveness, not least to different cetacean species themselves. To some degree, I wanted to resist the moralistic tendencies of postcolonial studies, the over-generalizations of posthumanism- or new materialism-inspired environmental humanities,[2] and the residual anthropocentrism of animal studies, although – inevitably again – I have probably ended up reproducing all of these deficiencies, each of which is deeply embedded in its own particular field. Perhaps most of all, I wanted to forage across the disciplines myself, with 'forage' being the operative word, that is collecting, bringing things together, sustaining oneself, or at least sustaining one's curiosity, from what one finds. Cetaceans can be curious creatures themselves, but I had no wish to conduct the kind of thought experiment that, say, Kathleen Jamie pursues in the above-mentioned Whale Hall in the Natural History Museum in Bergen, when, sitting directly under the vast skeleton of the blue whale, she follows its spine with her eye 'as it voyaged above the hall, curving very slightly, continuing between the other whales', and, extending her own sense of self in sympathy with it, tries to imagine 'what it might feel like to be a blue whale' (2012: 115).

Such acts of sympathetic imagination are always tempting, though as I will argue later they are simultaneously acts of *separation*, ironically reminding us that whales, however much DNA we might share with them, are nothing like ourselves. Still, I plead guilty to two things here: melancholy projection and an acute form of cetacean-related single-mindedness. Just as it helps to be melancholic to write about melancholy, it helps to be obsessive in writing about whales. One of the arguments I will pursue in this book is that while whales are substantial material presences, most of the time they are defined by their absence. Partially visible at best, whales remind us of what we can't see, and this lack of visibility allows us to better imagine what we have done to them. Another way of putting this is that whales function as multifaceted figures for human melancholy. While there are many different ways of understanding melancholy (see, for example, Radden 2000), for my purposes here I will define it as an afflictive condition, revolving around both real and imagined loss and both genuine and manufactured sadness, in which melancholy subjects produce the objects that they need in order to validate themselves (Pensky 2001 [1993]). In this context, the attribution of melancholy to the figure of the whale is less an effect of observable animal behaviour (ethology) than the projection of a human mood (sadness) together with an overriding guilt based on historical knowledge of the human exploitation of whales.

This mood can also produce violent anger, as it does in Ahab and several other celebrated examples, both fictional and real, of cetacean-centred single-mindedness. Whales are not by nature violent creatures, but violence is a standard commodity in the human narratives that surround them. As Ahab himself recognizes, melancholic loneliness is the flip side of this violence, born of the guilty recognition that the violence we have done to whales over the centuries is a violence we continue to visit upon ourselves. It is also a violence that haunts us. Philip Hoare's 2009 study *Leviathan*, one of the major inspirations for this book and inspired in turn by Melville, exudes this haunting quality. At the beginning of the book, Hoare summons up Melville's unquiet ghost, and in so doing brings to life the 'same creatures that came to obsess [his] ambiguous narrator' (39). (As Hoare suggests, *Moby-Dick* is in fact the product of *two* obsessions, that of its monomaniacal protagonist, Ahab, and that of its fatally compromised narrator, Ishmael – make that *three* obsessions if Melville himself is included as well.)

Like Hoare, in the four chapters that make up this book, I am embarking on 'an uncertain journey' in which I seek to 'discover why I too [feel] haunted by the whale' (39). There are personal reasons for this quest, but broader cultural reasons too, hence the book's tripartite title (*Colonialism, Culture, Whales*). As I have already suggested above, it is not possible to talk about whales without addressing significant issues of culture. Kalland's *Unveiling the Whale* (2009), one of few studies of its kind to adopt a comparative approach to the different cultures of whaling, usefully summarizes the different cultural arguments that have been used both for and against the whaling industry. Ellis (1999) further notes the minefield of (allegedly) cultural debates that continue to surround the viability or not of whaling practices, which encompass widely differing interpretations of subsistence needs, the role of local communities, and the protected status of hunting traditions, which now belong to a global cultural economy in which the definitions and regulations associated with whaling are still largely decided upon by the major pelagic whaling powers (see also Burnett 2012).

A test case for these debates is that of Makah whaling, which has attracted considerable critical as well as media attention. At least two different kinds of cultural arguments collide here. In one, as Marybeth Long Martello puts it, the Makah, 'like other aboriginal tribes in Canada and Alaska, maintain that whaling is necessary for preserving their identity and cultural heritage' (Martello 2004: 263). In the other, simplistic notions of 'traditional' indigenous cultures have been foisted upon them in order to make the case *against* whaling, on the grounds that the Makah, who now understandably adopt modern whaling practices, are effectively betraying their own cultural self-representation as 'natural' and 'authentic', with this view being based on an 'innate set of needs and abilities on the part of [members] of the tribe' (Martello 2004: 276; see also Russell 2010).

These cultural arguments came to the fore when in May 1999, the Makah shot and killed a grey whale just off the coast of their reservation in Washington State, arguing that they were perfectly entitled to do so, not so much in the interests of subsistence as of exercising their sovereign tribal rights. It soon became apparent that the case was more complicated than it had first seemed, with numerous variables in play such as different interpretations of tribal sovereignty, other forms of conflict within the community, and competing views, sometimes universally framed by anti-whaling activists, of environmental and animal rights (Russell 2010: 157). As Martello adds, the issue becomes more complicated still when one considers the various 'local' and 'global' arguments at stake. These include the need to protect 'local' culture over and against the 'global' imperative to protect cetaceans and other threatened animals, with globalization at once appearing as 'a defining limit of local cultures, a threat to these cultures, and a [broad] set of [environmental] obligations for traditional and indigenous peoples' in the context of an interconnected modern world (Martello 2004: 281).

There is no space here to tease these arguments out, but like Martello, I am interested in the broader questions they raise, such as 'what is culture in a globalizing world? How do we determine and mediate the needs of "local" culture and "global" nature? And who decides?' (Martello 2004: 280). *The Cetacean Quartet* no more has definitive answers to these questions than Martello does. But the questions themselves resonate through its linked discussions of contemporary whale watching as well as historical whale hunting, not least by challenging received wisdom that the one has effectively replaced the other, providing a low-impact, sustainable alternative that serves the 'global' conservationist cause while advancing various sets of 'local' commercial interests as well (Higham, Bejder, and Williams 2014). Here, too, cultural arguments can be brought to bear, which are often couched in terms of shoring up national interests. Thus, as the anthropologist Adrian Peace argues, whale watching in Australia 'has achieved considerable public prominence over the last decade or so because it resonates so effectively, not only with the image of Australia as a civilized society but also as a continent distinguished by an exceptional environment which can be exploited for tourist purposes yet kept fully intact and properly protected' (2005: 193). As Peace concludes, the mantra of 'sustainability' has allowed Australia to market itself worldwide as an environmentally friendly country that prides itself on the intelligent use of its own natural resources, despite considerable evidence that it does quite the opposite. Meanwhile, whale watching more specifically, even as it is touted as a paradigmatic example of such intelligent use, can be seen to be a flagrantly exploitative practice, which assumes 'that the worth of whales is to be determined according to our own consumption-fixated world-view alone' (2005: 206).

Peace's view is overstated, no doubt, but it points to what I will suggest in this book are the continuities between contemporary and historical whaling practices. Whale watching in and for itself may not be harmful to either people or whales, but a lot depends on the cultural and economic conditions surrounding it; and at worst, as in the notorious example of SeaWorld, it provides at least partial evidence that imperial relations are being re-established within today's global cultural economy through the empire-building capacities of transnational corporate power (Mignolo 2000; see also Chapter 3). As Peace and others suggest, the problem lies not just with the corporations themselves or the late capitalist system they support. It also lies with the public attitudes off which they feed, which seek sustenance in turn from a 'richly embellished anthropomorphic discourse' that complacently assumes the right to a personal relationship with whales (Peace 2005: 191; see also Higham, Bejder, and Williams 2014).

The irony that I am probably assuming this right as well is not lost on me here, and to some extent it is the occupational hazard of anyone who, partly following their own private interests, seeks to write about whales. It bears reminding though that whales are *not* like us, however much we try to incorporate them into our own explanatory narratives, and however much we search, by attempting to connect with them, for some kind of symbolic access to a lost or hidden past (Kalland 2009). The American historian of science D. Graham Burnett puts this well when he says that, unlike *primatology*, which is aimed at least in part at rethinking who we (humans) are, *cetology* has become a 'laboratory for the exploration of alterity, [...] a domain for thinking about who we are not' (Burnett 2012: 671). Whales inhabit a different realm, they see through sound, and in many – indeed most – conceivable ways they are profoundly indifferent to us. There are thus compound ironies behind what Burnett calls their 'oceanic quietism' – which is nothing other than an imagined escape from our own entrapment in self-made webs of 'ideology, alienation, materialism and violence' (Burnett 2012: 627) – and their symbolic instrumentality for the purposes of thinking through the 'common heritage of mankind' (Kalland 2009: 3).

The chapters that follow are designed in part to explore these ironies in the twin contexts of whale hunting (in the past) and whale watching (in the present). These contexts are seen in turn as capacious repositories of human–animal ideas, values, and attitudes which, though they may have changed significantly, by no means allow us to draw a line under our own violent past. Chapter 1 is organized around extinction themes, establishing both material and symbolic patterns – from the reality of exploitation to the spectre of extermination – that will be repeated throughout the rest of the book. Chapter 2 revisits some of the historical themes of Chapter 1, for example the beginnings and endings of whaling, looking more specifically at the pioneering role played by the late eighteenth- and early nineteenth-century northern English father–son combination, the Scoresbys, who were explorers and inventors as well as probably the best known whalers of their times. As in Chapter 1, I am interested in the roles of animal as much as human actors, so attention is paid, not just to the Scoresbys but also to the whales they assiduously studied (especially Scoresby Junior) as well as the ones they efficiently caught and killed. Chapter 3 then fast-forwards to the present to offer another look at a story that is as familiar now as the exploits of the Scoresbys would have been to their contemporaries: the trials and tribulations of the SeaWorld corporation, the world's biggest entertainment vehicle for the display of captive orcas (actually a species of dolphin, but popularly known as killer whales). This chapter is arranged around the relationship between killing and following, paying specific attention to the connections between social media and activism in both the 'outing' of SeaWorld and the creation of the celebrity orca, a repeatable figure that demonstrates the complicity between the pro- and anti-captivity narratives that circle around an industry which endangers the lives of the same animals it claims to protect. Finally, Chapter 4 returns to the scene of Chapter 1, both in a literal sense (the shores of Western Australia) and a figurative one (the spliced Ur-narratives of human/cetacean origins and human/cetacean extinction). The primary figure here is that of the melancholy whale, which, as also argued above, becomes a multipurpose catalyst for human-centred contemplations of loss and guilt as well as an opportunity for recovery and redemption. However, as befits the melancholic (*not* fatalistic) tone of the book in general, the conclusion reached is that many of these losses, both human and animal, will prove to be irrecoverable; thus, while saving the whales should remain an urgent priority for all of us, it may never be quite enough to save us from ourselves.

Notes

1 I am drawing indirectly here on Walter Mignolo's work, which posits 'coloniality' – his term – as the 'hidden face' of modernity (Mignolo 2000: 50). 'Coloniality' can be variously defined as a composite effect of colonial difference as the precondition for the subjugation of people and the 'subalternization of knowledges' (16); a site or platform for the building of alternative knowledges and narratives that emerge from local situations and histories; and a product of the recognition that modernity is not just European, nor is colonialism something that happens elsewhere than in Europe, rather both are part of a world system in which the colonial *constitutes* the modern and vice versa. The implications of Mignolo's views for both modern and traditional whaling practices are wide-ranging. First, it is generally accepted (and also argued in this book) that whaling is not just a product of, but a synecdoche for, Western industrial modernity and the imperial/colonial ambitions it spawned. Second, whaling helped to open up new colonial territories, which relied (quite heavily in some cases) on the whaling industry. Third and last, whaling – with its multinational crews and its increasing dependence on local forms of knowledge – frequently involved creolized and/or indigenous understandings of marine/coastal ecology, understandings linked to local cultural traditions, themselves often creolized, in which whales wielded considerable symbolic power.

2 Both posthumanism and new materialism are addressed in later chapters, so I will not offer working definitions here. However, it seems worth pointing out that neither term describes a critical movement as such, and that – much like environmental humanities in general – there are considerable differences in relation to both theory and practice. Not all posthumanists are new materialists and vice versa. Nor are these two internally conflicted approaches methodological

prerequisites for either postcolonial ecology or environmental humanities, both of which broad-based fields are perhaps more likely than, say, new materialism (though there are some notable exceptions) to focus on continuing global injustices and contemporary as well as historical differentials of power.

Works Cited

Burnett, D. Graham (2012) *The Sounding of the Whale: Science and Cetaceans in the Twentieth Century*. Chicago: University of Chicago Press.

Crosby, Alfred (2009) *Ecological Imperialism: The Biological Expansion of Europe, 900–1900*. 2nd ed. Cambridge: Cambridge University Press.

DeLoughrey, Elizabeth and George B. Handley (eds) (2011) *Postcolonial Ecologies: Literatures of the Environment*. New York: Oxford University Press.

DeLoughrey, Elizabeth, Jill Didur, and Anthony Carrigan (eds) (2015) *Global Ecologies and the Environmental Humanities: Postcolonial Approaches*. London: Routledge.

DeSombre, Elizabeth R. (2007) *The Global Environment and World Politics*. 2nd ed. London: Continuum.

Dodds, Klaus (2002) *Pink Ice: Britain and the South Atlantic Empire*. London: I.B. Tauris.

Ellis, Richard (1999) *Men & Whales*. Guilford, CT: Lyons Press.

Epstein, Charlotte (2005) *The Power of Words in International Relations: Birth of an Anti-Whaling Discourse*. Cambridge, MA: MIT Press.

Ghosh, Amitav (2005) [2004] *The Hungry Tide*. New York: Houghton Mifflin Press.

Haraway, Donna (2008) *When Species Meet*. Minneapolis: University of Minnesota Press.

Higham, James, Lars Bejder, and Rob Williams (eds) (2014) *Whale-watching: Sustainable Tourism and Ecological Management*. Cambridge: Cambridge University Press.

Hoare, Philip (2008) 'Troubled Waters: Did We Really Save the Whale?' *The Independent*, 19 September www.independent.co.uk/environment/nature/troubled-waters-did-we-really-save-the-whale-935193.html, accessed 25 June 2017.

Hoare, Philip (2009) *Leviathan or, the whale*. London: Fourth Estate.

Hogan, Linda (2009) [2008] *People of the Whale*. New York: Norton.

Howe, Stephen (2002) *Empire: A Very Short Introduction*. Oxford: Oxford University Press.

Ihimaera, Witi (2005) [1987] *The Whale Rider*. Rosedale, NZ: Penguin Books.

Jamie, Kathleen (2012) *Sightlines*. London: Sort of Books.

Kalland, Arne (2009) *Unveiling the Whale: Discourses on Whales and Whaling*. Oxford: Berghahn.

Martello, Marybeth Long (2004) 'Negotiating Global Nature and Local Culture: The Case of Makah Whaling'. In S. Jasanoff and M.L. Martello (eds), *Earthly Politics: Local and Global in Environmental Governance*. Cambridge, MA: MIT Press, 263–284.

McGuire, Ian (2016) *The North Water*. London: Scribner.

Mda, Zakes (2005) *The Whale Caller*. New York: Viking.

Mignolo, Walter D. (2000) *Local Histories/Global Designs: Coloniality, Subaltern Knowledge, and Border Thinking*. Princeton, NJ: Princeton University Press.

Peace, Adrian (2005) 'Loving Leviathan: The Discourse of Whale-watching in Australian Ecotourism'. In J. Knight (ed.) *Animals in Person: Cultural Perspectives on Human-Animal Intimacy*. Oxford: Berg, 191–210.

Pensky, Max (2001) [1993] *Melancholy Dialectics: Walter Benjamin and the Play of Mourning*. Amherst, MA: University of Massachusetts Press.

Radden, Jennifer (ed.) (2000) *The Nature of Melancholy: From Aristotle to Kristeva*. New York: Oxford University Press.

Russell, Caskey (2010) 'Wild Madness: The Makah Whale Hunt and Its Aftermath'. In B. Roos and A. Hunt (eds) *Postcolonial Green: Environmental Politics and World Narratives*. Charlottesville, VA: University of Virginia Press, 157–176.

Sciullo, Nick J. (2008) 'A Whale of a Tale: Post-Colonialism, Critical Theory, and Deconstruction: Revisiting the International Convention for the Regulation of Whaling through a Socio-Legal Perspective'. *The City University of New York Law Review* 12.1: 1–27.

Shukin, Nicole (2009) *Animal Capital: Rendering Life in Biopolitical Times*. Minneapolis: University of Minnesota Press.

Steinwand, Jonathan (2011) 'What the Whales Would Tell Us: Cetacean Communication in Novels by Witi Ihimaera, Linda Hogan, Zakes Mda, and Amitav Ghosh'. In E. DeLoughrey and G.B. Handley (eds) *Postcolonial Ecologies: Literatures of the Environment*. New York: Oxford University Press, 182–199.

Vizenor, Gerald (1999) *Manifest Manners: Narratives on Postindian Survivance*. Lincoln: University of Nebraska Press.

1

Last Whales: Eschatology, Extinction, and the Cetacean Imaginary in Winton and Pash

Ah the world, oh the whale.

Philip Hoare, after Herman Melville

The great majority of interpretations of Apocalypse assume that the End is pretty near. Consequently the historical allegory is always having to be revised; time discredits it.

Frank Kermode

Introduction: Soundings

Early in 2015, the management team of the Natural History Museum in London made a momentous decision. A hugely popular diplodocus cast ('Dippy') would be replaced in the museum's central hall by the full-size skeleton of a blue whale, which had previously been on display in the animals' gallery, having first been acquired as one of the museum's earliest specimens in 1891. The decision to give pride of place to a whale was justified on conservationist grounds; as Sir Michael Dixon, the museum director, explained: 'The very resources on which modern society relies are under threat. Species and ecosystems are being destroyed faster than we can describe them or even understand their significance. The blue whale serves as a poignant reminder that while abundance is no guarantee of survival, through our choices we can make a real difference. There is hope' (quoted in Kennedy 2015: 23).

This latest instance of the whale as conservationist icon – as outsize standard-bearer for marine and other environmental issues – serves as further confirmation of whales' immediate emotional appeal as well as their enduring symbolic power (Burnett 2012; Hoare 2009). As the British writer Philip Hoare puts it, no other creature has the capacity to 'represent life on such a [grand] scale or to act as an antidote to our own constricted [existences]', while no other seems able to capture the popular imagination quite like the whale, which is both like us and profoundly unlike us, which straddles myth and reality, and which has latterly been mobilized as a symbol of lost innocence in an age of global environmental decline (Hoare 2009: 29–30, 32–33).

Listed as endangered today, the blue whale also serves as a graphic warning that whales and other cetaceans, though only some species among them are currently considered as being at serious threat, have historically operated under the sign of *extinction* – and not just the partly realized possibility of their own disappearance but also that of humanity and even the planet itself. A hallucinatory vocabulary of termination surrounds the whale – a quickly spreading 'ecology of fear' (Davis 1999) that is at times expressed in broad religious terms, at others refracted through more specific environmentalist anxieties. For whales are not only worlds unto themselves, but also ways of thinking about the world and human beings' increasingly precarious place within it (Burnett 2012: 329). Today, whales are at once powerful symbols of the fight to halt the continuing destruction of the planet and all-too-material reminders of the legacies of that destruction. Once routinely viewed as harbingers of apocalypse (Hoare 2009: 60; see also Roman 2006), they are now more likely to be allegorized as promissory vessels of human redemption whose protection is urgently needed in order to save us from ourselves (Burnett 2012: 329).

There seems no end to the ways in which whales can be turned, whether literally or metaphorically, into something *else* even as they are now increasingly acknowledged as enjoying special status among the earth's creatures, as being ecological subjects in their own right (Kalland 2012). It is not so much that whales are allegorized, rather that they are *obsessively* allegorized; not so much that they are anthropomorphized, but *relentlessly* anthropomorphized, for example in terms of demonstrating the qualities – kindness, caring, compassion – we humans would like, but usually fail, to see in ourselves (Kalland 2012: 41). Much of the hyperbole has to do with time. Whales straddle markedly different versions of the past, bringing together the *longue durée* of capitalist modernity (the all-too-human basis for whaling history) with the still longer stretch of an age that both predates and dwarfs human presence (the other-than-human reaches of prehistoric time). The hyperbole also has to do with the *lack* of time: with the perception that time is fast running out, and that whales consequently represent a kind of planetary last chance in the face of impending catastrophe. As Sir Peter Scott, the then chairman of WWF, would say in 1972 in the early days of what would go on to become the global 'Save the Whales' movement: 'The feeling is now abroad that if we can't save the largest animals we know we have little chance of saving the biosphere itself and therefore of saving our own species' (quoted in Simmonds and Hutchinson 1996: 465).

While much has been done since then to protect the whale, fears surrounding it persist, many of them bound up with apocalyptic forebodings of different kinds that, together, amount to a full-blown 'planetary dysphoria' in which new and inventive ways are currently being imagined of how terminal ecological catastrophe might be visited upon a sentient earth (Apter 2013: 335). One of the many ironies of planetary dysphoria is that it has given a new lease of life to death, which can now be seen in the large and growing body of contemporary thanatological writings that seek either to think through the consequences of planetary destruction or, going beyond it, to contemplate the hypothetical extinction of thought itself (Apter 2013; Brassier 2007; Colebrook 2014a and b; Morton 2013).[1]

In some of these writings, a working distinction is posited between *eschatology*, which anticipates the inexorable termination but also the eventual revitalization of life, and *extinction*, in which such renewal is no longer possible (Brassier 2007). The distinction is specious, however, insofar as eschatology and extinction, despite the aura of finality hovering over both, are temporally complex. Certainly, eschatology – understood here as the Judaeo-Christian doctrine of last things – *can* be seen in teleological terms, as the linear process by which 'things reach their end when they fulfil the purpose for which God created them' (Walls 2008: 4). But theologians are not necessarily agreed on this point, with some interpreting eschatological time as linear and singular and others as cyclical and periodic, while still others argue over whether the foreseen end is imminent or might be indefinitely deferred (Schwarz 2000: 27).[2]

Nor is extinction unequivocal. Thus, while extinction is sometimes popularly framed as an abrupt end event, it is more often than not subject to longer processes that involve the gradual foreshortening of not just a single *life form*, but multiple and multiply interrelated *forms of life* (Van Dooren 2014: 5; see also Kolbert 2013). Extinction, argues the Australian environmental scholar Thom van Dooren, can happen in any number of different ways that yield any number of different meanings, but one constant is that it is not just the death of the last representative individual of a given species, but is part of a patterned process in which one loss involves others – often *many* others – over an extended period of time. Van Dooren duly points to what he calls 'the dull edge of extinction' (12), in which there is 'a slow unravelling of intimately entangled ways of life that begins long before the death of the last individual and continues to ripple forward long afterward, drawing in living beings in a range of different ways' (12).

This chapter is a preliminary attempt to gauge the different temporalities that attend the ambivalent figure of the whale as an index of *both* salvation *and* perdition in the dual context of eschatological readings of planetary destruction and environmentalist readings of ecological collapse. In many such readings, whales are spectral figures moving almost imperceptibly between alternate temporal registers: now representing humanity's hopes for its own future, now burdened with the collective deadweight of humanity's destructive past (Lippit 2010).[3] Extinction throws up another possibility, however: that of thinking *beyond* the human social relations for which the figure of the whale is often metaphorically substituted, and *beyond* the natural world for which it often metonymically stands (Kalland 2012: 3). Steadfastly anthropocentric projections of the kind Kalland references here raise the question, not so much as to whether humanity is worth saving in the first place, rather to what extent 'the human' has outlived its usefulness as an ontological category and needs to make way for alternative modes of conceiving planetary co-existence that are not necessarily dependent on the survival of human beings in their own meticulously manufactured world (Colebrook 2014a).

The two perhaps unlikely texts I want to turn to in order to explore these different extinction scenarios are Tim Winton's 1985 novel *Shallows* and Chris Pash's 2008 journalistic account *The Last Whale*. Both books are written by conservation-conscious Australians for whom the environmentalist stand-offs that led to the late twentieth-century abandonment of commercial whaling in Australia – their ostensible subject – are explicitly or implicitly measured against a plurality of time scales in which human history either destructively repeats itself or is contained within a much longer natural history that ultimately proves beyond human capacity to comprehend. Against these broader metaphysical backgrounds, which are framed by Winton in eschatological terms, competing micro-histories of extinction are played out involving a wide variety of both human and non-human actors. In both texts, as will be seen below, whales are more than just symbols, registering a substantial material presence. In both texts, though, whales are also reversible signs within a symbolic economy in which 'religious' as well as 'scientific' interpretations founder (Winton), and in which regional, national, and international gesture politics – with all the familiar human failings that accompany them – risk hijacking the broader moral imperative to protect non-human life (Pash).

Sightings (I): The maw of History

The 1985 novel *Shallows* is one of several of Winton's fictional works to be set in Angelus, a thinly disguised if still densely mythologized version of Albany, the small but symbolically significant Western Australian fishing port, one of whose claims to fame derives from being both a *first* and a *last* place, doubling as a remote colonial outpost and as the region's earliest settled town (Winton 1985: xi). As is axiomatic for Winton's work, Angelus is ironically named within a tragicomic eschatological context. Deriving from the Latin for angel, 'angelus' is the first word of a Catholic recitative to the Virgin Mary in praise of the Incarnation, an incarnation that is also transfiguratively embodied in the seasonal humpbacks which, Christ-like, are ritually slaughtered and then symbolically resurrected through their annual return to the bay (Thomas 2008: 18).

For Angelus is also a historic whaling town, one of the key sites for colonial Australia's first major industry (Turner 1993: 79).[4] Appropriately enough, the first whale to be seen in the novel is stone dead, a rotting humpback 'jettisoned by the sea [and] left alone by even the sharp-beaked gulls that hunt the lonely shallows for smelt and mullet' (Winton 1985: xii). The year is 1831, and nearby lies another 'wasted hulk' (xii), this time in human form, namely the half-starved American whaler Nathaniel Coupar. Coupar is a now largely forgotten founding figure whose Australianized ancestors are the unlikely heroes of Winton's novel, most of which is set in the late 1970s, at the time Australia's last land-based whaling station (at Frenchman Bay near Albany) was forced, in the wake of a seemingly unstoppable wave of local and international environmentalist protest, to shut down (Turner 1993: 79–80).

The novel plays fast and loose with these and other historical facts, operating in the spirit of a fashionable sub-genre at the time, 'historiographic metafiction', in which history is energetically retold as fiction and legend, the uneven power relations underlying such twice-told tales are critically examined, and both history and fiction are enjoined to reflect recursively on themselves (Hutcheon 1989). (In this and other respects, *Shallows* is very much a novel of its day, readily connected to late twentieth-century debates around the aesthetics and politics of postmodernism; but as I will seek to argue in this chapter, it moves between different temporal registers and is just as easily assimilated to early twenty-first-century ecocritical discussions of distributed agency in a more-than-human world.) Also in the spirit of historiographic metafiction, *Shallows* pieces together its accounts from the self-duplicating fragments of different narratives, with the Bible – pre-eminent but by no means definitive among these – proving no more reliable than Nathaniel Coupar's conscience-stricken diaries, and no more foundational than their mid-nineteenth-century counterpart, that legendary Great Whale of a novel, itself a loose compendium of founding narratives, *Moby-Dick*.[5] One effect of this plethora of uncanny re-doublings is to restage authoritative accounts as entertaining yarns, which itself has the dual consequence of dismantling local whaling history and of upsetting the supposedly universal narratives that underlie it, with History, Science, Religion, and other grand narratives all being downsized into so many boastfully embellished fishermen's tall tales.

One prime example is the story of Jonah and the Whale, which is retold several times during the course of the novel (88, 124, 197; see also Chapter 4). Jonah and the Whale is usually interpreted as a classic Old Testament parable of repentance and redemption, with Jonah eventually being spared after praying for forgiveness, and the whale releasing him to complete his God-given task of rescuing the world from its fallen state (Roman 2006: 10–12). But as another of the novel's stalwarts, Protestant minister William Pell, says in recalling some of the old tales he used to exchange with Nathaniel's grandson, Daniel Coupar:

> I remember that one about Jonah – oh, there were a dozen variants – where Jonah is swallowed by the whale and in the whale he meets the Devil himself and the fight is on. It's the violence of the fighting that makes the creature spew Jonah back up on the beach. A neat little ending with the Evil One dragged off into the deep, still a captive. And the language of the fight! I never heard such filth in all my life – the things Satan said to Jonah and the words Jonah chucked back – and I've never heard them used so well since. I can't think what the masters would have done to him [Coupar] if they'd heard. Ah, he was a wonderful liar … Always used the truth when it suited him though. And I s'pose the truth always used him when it suited itself. (197)

Here, the whale becomes a carrier of truths of which it (and possibly Jonah himself) is unaware, as well as a vehicle of falsehood; it doubles as God's messenger and as Satan's agent, bearing both of these identities hidden *within* rather than inscribed *upon* its body, as if to show the fundamental uncertainty of God's grace (13, 16).

The whale is a reversible sign, then, the interpretation of which is always liable to shift between linked extremes, just as the angelic always risks sliding into the demonic (123). And at another, still more general level, it is condemned to be *nothing but* a sign, to always be something other than itself. In this last sense, *Shallows* can be seen as relating an epic struggle over the different meanings, both spiritual and not, that have historically attached to the cetacean imaginary: meanings that pose considerable difficulties of navigation, just as the 'real' whales of the text seem unable to decipher the consequences of their own movements, and eventually strand themselves. How are whales to be 'seen' in the text, and what happens when they are 'sighted'? What happens when they are heard but not seen, or when they are not seen clearly? What happens when they *are* seen clearly, but are still taken to be signs of other, absent presences, summoning up spectral visions from the past? And what happens when, Jonah-style, they are seen as little more than gaping mouths, laid traps for the foolish and unwary, or as gargantuan embodiments of the maw of History, into which the virtuous and the vicious alike are inexorably absorbed?

The novel explores all of these possibilities and more, connecting each to an eschatological frame of reference that works according to the principle of anterior prefiguration: for it is not just that the world the novel presents is *fore*told as dying, but *re*told as if it were already dead. Angelus, to put this differently, is in a state of terminal decay, less a place of refuge than refuse, unwanted residue from the shiftless present as well as the convict past (6). It is a place of accumulated 'last things', increasingly derelict home to a motley crew of drifters and deserters, wrecks and reprobates; even the drought-plagued land itself, pinched coastal territory caught between a huge sea and an arid inland, is described as 'putrefied and pussed up [*sic*] like a dried scab' (14). Perhaps more to the point, it is a place of *remnants*, men in their last days, surviving but with death already marked upon them; and a place of *corpses*, stinking bodies alternately human and non-human, each superimposed onto the other like so many 'wasted hulks' (xii).

In this and several other ways, *Shallows*, like so much of Winton's work, exercises a fascination for the macabre, which is not the same thing of course as a weakness for the morbid; rather the macabre demonstrates a heightened *consciousness* of the morbid, which deliberately exploits it for comic effect. Much of *Shallows* in fact, despite its hauntingly melancholic tone, reads as mischievous black comedy. And much of it functions as a kind of *mock-eschatology* that implicitly ridicules the same endings it foreshadows, playfully riffing on the rhetoric of extinction (significantly, Angelus's two warring family figureheads, Daniel Coupar and Des Pustling, are both presented as being the 'last of their line' [74, 159]). Few of Winton's critics seem to want to acknowledge this, which is a particular problem for those religious-minded commentators who see his fictional work as a more or less transparent attempt to articulate the numinous with the ordinary and to communicate, albeit unconventionally, the mysterious workings of God's grace (see, for example, Miels 1993; Thomas 2008; Turner 1993).

Whether Winton is a 'religious' writer or not is hardly the point; rather what interests him is the extent to which alternative explanations of existence tend to be betrayed by the very language they seek to manipulate in their cause. A case in point is *Shallows*' lively parody of the 1970s' language of radical environmentalism, which is repeatedly exposed in the text as either self-serving – 'SUPPORT SALLY MILES/REDEEM YOURSELF: SAVE THE WHALES', reads one protestor's placard (145) – or marked by New Age clichés which frame the cetacean imaginary in terms of its opposite, the failure to conceive of whales in other than the most familiar and achingly sentimental terms. ('The whales have become my life. They are the most amazing creatures alive. They have intelligence, wit, compassion. There is much that is mystical about the whale. One day, if we keep them alive long enough, we will discover it, and perhaps learn something about ourselves. They are almost magic, the friendly giants of our childhood dreams. Think of the things the whale has seen, the civilizations coming and going [...]. They harbour secrets. I want Man to know them one day' [156].)

This last quote is from Georges Fleurier, one of the founder members of Cachalot & Company, the ramshackle activist group that comes to Angelus with the explicit aim of closing down the Paris Bay Company and thereby ending commercial whaling in Australia – next stop the world. Loosely based on the flamboyant French Algerian entrepreneur Jean-Paul Fortom-Gouin, aptly described by Chris Pash as an 'activist in a business suit' (Pash 2008: 21), Fleurier succeeds in alienating half of Angelus with his exhibitionist antics, some of which are likened to those of Ted Baer, the celebrity American 'shark hunter' who has come to Angelus in search of the huge white pointers its coastal waters are famed for – and in order to attract further attention to himself. To his credit, Baer makes little attempt to be other than shallow; Fleurier manages almost effortlessly to be shallow while pretending to be deep.[6] Slightly better, though still unprepossessing, is another American, Marks, whose main role for Cachalot & Company is to expatiate on whales as marvels of science while claiming the need for their protection irrespective of the scientific theories that surround them: 'A thing doesn't need to be intelligent', he says, 'to need a reason to be' (152–153).

Given Winton's own activist credentials, it is difficult to know what to make of this.[7] Clearly he is poking fun at fledgling environmentalist groups that have been more derisively described by D. Graham Burnett as 'a community of grandstanding saboteurs and dope-addled visionaries' (2012: 528). With Winton as with whales, it can be hard to navigate the signs, to gauge the text's continual slippage between the literal and the metaphorical in order to recognize which of these registers one is in (it is often both). Certainly, *Shallows* is a text that resists its own moralizing tendencies, but it would be unwise to conclude from this that Winton is breezily dismissive of all kinds of environmentalist protest or necessarily opposed to the more theatrical forms of activism practised by radical environmentalist organizations. In its characteristically side-on way, *Shallows* memorializes the people it mocks, just as it commemorates the history from which it seeks to deliver itself. But equally characteristic is the darkness in the text, which fixates, Ahab-like, on the very subject that is most likely to destroy it. In rejecting the sentimentalism of the 'friendly' whale, *Shallows* casts a shadow over one of the earth's most consistently exploited creatures, which still has the capacity to inhabit our worst nightmares – take Nathaniel Coupar's diaries, which conclude with their author's suicide in a desolate coastal landscape in which there is little hope of rescue or redemption, and 'rotting caverns of bone lay in the still shallows near the shore' (171).[8]

Sightings (II): The last of its kind

Chris Pash's *The Last Whale* (2008) can be read in some respects as a non-fictional version of *Shallows*, drawing on previous material collected at the time for the *Albany Advertiser* on the breathless sequence of regional, national, and international events that would eventually lead to the closure of the Cheynes Beach whaling station in 1977. The book is endorsed by Winton but dedicated to Greenpeace, whose involvement in the 1977 protests was its first Australia-based direct action campaign. A fairly standard journalist's account, *The Last Whale* is short on artistic flair but technically accurate and even-handed, showing awareness of the skills involved in hunting whales as well as campaigning against the hunt, and – though clearly conservation-oriented – taking care to see the relevant issues from all sides.

The text also captures the complexities involved in conservation, including the tactical battles fought out not just over the body but, even more crucially, the *image* of whales. 'The whale was a symbol and yet they were killing the whales', complains one of the main activists involved, the spirited but less-than-sophisticated American Aline Cheney, about the whaling industry in Albany: 'They loved the image [evident in Albany's branding as a "whale town"] but were killing the real thing' (87). The campaign to close Cheynes Beach duly emerges as a heady combination of business savvy and media glitz, explicitly designed to create a 'domino effect' (27) that might eventually extinguish global whaling. However, the theory isn't necessarily matched by the practice, which often relies on gimmickry (Willy the life-size blow-up whale) and thuggery (the Japanese are easy targets), and which just as frequently betrays the amateurism of its assorted international players, who are simultaneously presented as naïve and calculating, clueless and clued-up (117).

The 'whale war' (141) is thus waged on several different fronts, with grassroots environmentalism, federal politics, and local livelihoods all being caught up in the tangle (184). Another way of putting this is that there are multiple extinctions at play organized around the symbolic figure of the whale: the extinction of the whales themselves (92), the extinction of the whaling industry (27), and the extinction of whaling as a local way of life (188). Looming over all of these is a further possibility, the eventual extinction of human beings. As another of the activists, local Australian Tom Barber (who will eventually marry Aline Cheney), speculates, drawing on the ideas of the American systems theorist Buckminster Fuller:

> [W]hat could an individual do to change the world and benefit humanity? The big question was whether humanity would survive in the long term. What the world needed was continual great leaps to counter the degradation of the planet. Tom thought that energy from the wind – a gift from the sun's gravity – was the way to go. (124)

As so often in the text, this isn't really thought through, and it is left to the more experienced Canadian activists, Greenpeace co-founders Bob and Bobbi Hunter, to work more closely with the media to 'beam [consciousness-changing] stories and images [of] dead and dying whales into the homes of the people of the world' (112). Pash, too, participates in this media blitz, and while his account is for the most part understated, he is certainly not averse to ventriloquizing the emotive vocabulary of extinction, as when Bobbi Hunter encounters the Cheynes Beach whaling station for the first time:

> The [...] station was a place of grisly horrors for her, the flesh-stripped bones of the whales a monument to genocide. Images of how they died flooded her mind and she cried. But she looked across at Bob, who sent a thumbs-up, and she knew it was okay to cry on national television. (92)

Later in the text, the threat of extinction is used as a basis for the national conservation effort, led by the newly re-elected prime minister, Malcolm Fraser, one of whose first actions in office is to announce an independent judicial inquiry into the whaling industry in Australia. Pash reproduces Fraser's words, which in turn are given in full here:

> Many thousands of Australians – and men, women and children throughout the world – have long felt deep concern about the activities of whalers. There is a natural community disquiet about any activity that threatens the extinction of any animal species. I abhor such activity – particularly when it is directed against a species as special and as intelligent as the whale. There are, however, two distinct views in relation to the activities of whalers. One view put to me strongly is that all whale species under threat of extinction are protected by moratoriums imposed by the International Whaling Commission and that current policy is in line with the best principles of conservation. An alternative view, which has also been strongly argued to me, is that the present practice of killing whales does endanger the whale species. Many other arguments have been put on both sides. (160)

The second set of arguments eventually wins out, aided and abetted by whalers' unconvincing accounts of 'humane killing' as well as more dispassionate commercial factors such as rapidly falling oil prices and other associated global market trends (170–171; 190). This all leads to the crowning image of the text, which is that of the 'last whale' caught by the Cheynes Beach Whaling

Company, described as a carnival affair in which, despite impending job losses, the whalers dress up, their vessels fly flags, and all concerned do what they can to enjoy themselves (191). The whales themselves may not be extinct, but the occasion, despite the atmosphere of forced hilarity it creates, melancholically marks the 'dull edge of extinction' (Van Dooren 2014: 12) insofar as only one life is lost, but many others – human as well as non-human – are understood to be involved. Still, Albany's loss is Australia's gain, with the country rapidly evolving into a 'global advocate for an end to whaling' (Pash, 2008: 195) – a good example, this last, of what Arne Kalland calls the 'symbol-politics' played out by former whaling nations now jostling with one another to 'appear to be civilized states that care for nature' on the global stage (2012: 193; see also Chapter 4).

The ironies of this are not lost on Pash, whose epilogue to *The Last Whale* includes a telling entry from Project Jonah financier and Greenpeace Australia co-founder, Richard Jones. The Cheynes Beach campaign, brags Jones, 'should serve as a blueprint on how to win an issue. It not only stopped Australia whaling, but turned Australia from being one of the most fervent pro-whaling countries into the most anti-whaling country' (201). This goes uncommented, but reading between the lines, it is possible to detect a tacit criticism of national gesture politics – gestures that are also liberally reproduced at regional and international levels in the text. In each case, the whale serves as a 'totem animal' (209), image rights over which are linked to several different ethical commodities – personal responsibility, national conscience, human survival – and played out on several different political fronts. *The Last Whale* thus quietly points to the political game playing that has continued to accompany anti-whaling campaigns in Australia and other Western countries, campaigns which have turned from national to global, but with the broad moral imperatives underscoring them hiding a variety of more local – still often national – concerns.[9]

Central to many of these campaigns is the idea of whales as creatures of 'special significance' (Pash 2008: 203), a phrase that resonates across a wide variety of late twentieth- and early twenty-first-century whaling literature, but that serves an equally wide variety of interests involving what might best be described as different philosophical and political repurposings of the whale. In his magisterial 2012 study *The Sounding of the Whale*, D. Graham Burnett describes the seemingly improbable process by which whales, biologically classified as 'an anomalous order of elusive, air-breathing mammals', came politically to serve as 'nothing less than a way of thinking about our planet', giving rise to a 'new creature of extraordinary symbolic power' (643) forged from the fragments of old myths as well as more recent scientific discoveries, and marrying misty New Age spiritual aspirations to more clear-headed (if not always clearly articulated) environmental concerns (626).

This, broadly speaking, is the 'new whale' of Winton and Pash, freshly fashioned from the late twentieth-century mesh of postcolonial politics and New Age environmentalism. But this 'new whale' doesn't necessarily have new habits; on the contrary, it simultaneously occupies a number of different time frames, in only some of which it is 'new' and in far more of which it precariously survives as a historical remnant (Winton) or, metaphorically at least, as the last of its kind (Pash). What is more, the nature of this survival is unclear. In both texts – albeit more explicitly in Winton's – whales are transitional creatures of *eschatological time*, which is itself split between a 'time of salvation' and a 'time of crisis' (Schwarz 2000: 27), and in which both of these temporalities take on ecological dimensions that, depending on perspective, locate the time of salvation either in an anticipated future or as *already past*. For most Christian theologians, contemporary environmental concerns belong to the 'time of crisis'. Hans Schwarz, for instance, sees 'the anticipatory power of eschatology [as giving us] gives us the incentive for stopping the exploitation of our environment and for preventing our own self-destruction' (204), although he does not rule out the less attractive possibility that 'through our neglect and wanton destruction we might involuntarily bring the *eschaton* upon ourselves' (196). For Christopher Partridge, however, the 'eschatological scenarios' conjured up by many contemporary (especially New Age) environmental movements betray despair of 'a political or religious transformation of the world, [envisaging instead] cataclysmic intervention of a divine, otherworldly, or superhuman kind' (Partridge 2008: 192). In some of these scenarios, there is a fear that things have been left too late: that the possibility of salvation no longer exists or, if it does, that its time has already effectively passed. Eschatological time, in this apocalyptic mode, gives way to a time of extinction that instantiates the slow but inexorable 'unravelling of intimately entangled forms of life' (Van Dooren 2014: 12). *Stranding* – the last actions of last whales – is the motif that best captures the ambivalent nature of eschatological time as it applies to cetacean species, and to a cetacean imaginary in which whales are seen as morally freighted portents of an uncertain future as well as 'supernaturally physical' remnants of a destructive past (Hoare 2009: 30).

Conclusion: Strandings

It still remains unknown what causes mass whale strandings, although the scientific literature lists a number of possible contributing factors that range from 'a Malthusian instinct for the preservation of the species' to navigational difficulties to the workings of genetic memory in relation to the evolutionary past (Hoare 2009: 304). (Climate change is another likely factor, as appears to be the case in several recent mass strandings.) Strandings are in turn baffling and disturbing for those who witness them. Beached whales have historically been taken for gifts from the gods, but also interpreted as evil omens (Hoare 2009: 305). Depending on which of these interpretations is favoured, some indigenous cultures have lured whales into the shallows in a kind of welcome ceremony (Roman 2006: 38), while others now weep at the sight, doing all they can to free the victims, which operate as they once did in medieval Europe as 'signs that all is not right with the world' (Roman 2006: 179).

The motif of stranding in *Shallows* is similarly Janus-faced, interpreted by some (e.g. Queenie Cookson) as a kind of suicide without redemption, a futile act of self-extinction, and by others (e.g. Georges Fleurier) as a complex sacrificial ritual: 'Death is a testament', he says in conversation with Queenie, 'One can learn a great deal about the life and mind of something from the manner in which it dies' (Winton 1985: 47). While Fleurier appears – for once – to be right, it is moot how much if anything is learned, either more generally from Winton's cetacean fable or more specifically from the nocturnal scene of mass stranding, which is as appalling as it is predictable, with which it ends. The circumstances are worth recounting. Queenie and her estranged husband Cleve, who have recently been reconciled, are woken by the familiar sounds of humpback whales, joyfully interpreting these as the sign of a return to normality; for Angelus has been anything but normal for a good while now, and earlier whale sightings, while still occasionally touched with foreboding, had held out the promise of a return to seasonal rhythms as well as a reabsorption into the older – ancient – histories connected to ancestral memory and deep time (13, 79). But this latest return has nothing reassuring about it:

> Rapid scalar movements, changes of tone, sounds of unmistakable emotion came to them, and the Cooksons dressed and rushed outside with a torch and ran down the wet sand in the rain and shone the torch and saw the huge, stricken bodies lurching in the shallows. Queenie screamed. Surf thundered and the night was images in torch beams. Masses of flesh and barnacles covered the sand, creeping up, floundering, suffocating under their own weight. A pink vapour from spiracles descended upon Cleve and Queenie as they moved between the heaving monuments. (290)

This disastrous failure of navigation – if that is what it is – is thrown upon the reader, who struggles as Queenie and Cleve do to make sense of self-destructive actions that seem, at one and the same time, so dense with meaning yet so incomprehensible to those who witness them and are seemingly as powerless to stop them as they are to intuit their cause. Winton's critics have struggled too, with John P. Turner, who initially seems to favour a religious reading based on the apocalyptic return of the Leviathan, eventually precipitated back into a series of empty rhetorical questions: 'Where is the release from suffering and the escape from guilt to be found?' (1993: 85). As Turner lamely concludes, 'the final, climactic moments of the book are ambiguous ones, with all the complexity of Christian eschatology' (85). This is true, but misses the point that 'religion' no more provides an adequate explanation than does 'science' for some of the things that happen in Winton's novel. Compare the following, detailed explanation of stranding from Marks, the novel's self-designated 'strandings expert':

> There's a heap of theories, you know [Marks explains to Queenie]. The echolocation faults are the most popular at the moment, like the whaletrap theory. You see, when a pod of whales is moving north along a complex coast with inlets and coves and deep bays they sometimes come into a bay which is so deep and big with a sweeping headland that in order to get out again they might have to swim south for a distance. You know, exactly in the opposite direction their migratory senses tell them to. Their whole beings compel them to move north – to escape they must move south. They get distressed,

> hesitate long enough in the swell which is often heavy – or the tide – and they can get caught in the shallows [...]. Whales don't operate their best in shallow water. Very flat, long, shelving beaches are their traps. The water is warm in the shallows – they like it – but their sonar gets hazy in that kind of uniform terrain. They can't identify it properly, make mistakes, get frantic; they're stuck. And all this complicated loyalty. If one goes, all go. (150–151)

This is fine as far as it goes, but like much of the rest of the text it leaves the reader wondering how much of its language should be treated metaphorically; for it seems clear that one of the fundamental problems both *in* the novel and for a reader *of* the novel is how to go about interpreting signs (Thomas 2008). The whale is itself a sign, and an arbitrary one at that, which makes for all kinds of interpretative difficulties; as Daniel Coupar recalls from Angelus's wartime years, soldiers on the look-out for submarines would sometimes breathlessly claim to have confirmed a sighting, only to have it reconfirmed as a whale (Winton 1985: 78). If, on another level, the whale is much more than a sign, the problem is that it has continued to be interpreted allegorically, even as its massive body has itself been turned into an allegorical ruin, raw material for something else (Hoare 2009).[10]

Times of crisis usually have the effect of multiplying signs: hence the hyperactive semiosis of *Shallows*, which draws to some extent, as Winton himself acknowledges, on the even richer semiotic resources of *Moby-Dick*. The more-than-human[11] corpses and carcasses that litter Winton's text function as both remains and reminders, as temporal signs – deictic markers – whose meanings shift from one location to another, depending on each new temporal context in which they are found. Eschatology cannot solve this semantic problem; it can only demonstrate it, shifting as eschatological thought does between alternative temporal registers in which finality is always over the horizon, either situated in some yet-to-be-decided future or embedded in some yet-to-be-acknowledged past. Even extinction cannot solve the problem insofar as it draws attention to the multiple futures it forecloses, each subtended by an evolving species that, by definition, is 'always becoming different from, other than, itself' (Van Dooren 2014: 38). This is not fatalism any more than even the most melancholic strands of contemporary extinction theory imply a meek abandonment to the various ends and endings they imagine – to the accumulated 'tomorrows of non-existence', to borrow Claire Colebrook's felicitous phrase (2014a: 43). Rather, as

Colebrook suggests (2014b: 148), it is a way of conceiving of a world that is no longer defined by the human will to survive: a world in which whale strandings – whales themselves – are not deathly portents, but complex manifestations of a life that co-exists with humanity, but is not reducible to human understandings of history, still less to the possible futures or non-futures that human beings might imagine for themselves.

Notes

1 As Claire Colebrook argues, there are three senses of extinction: the sixth great extinction event, the extinction by humans of other species, and self-extinction, which she defines in terms of 'the capacity for us to destroy what makes us human' (2014a: 9). Like several other contemporary theorists of extinction, Colebrook links self-extinction to the category of the *posthuman*, which in its most extreme – some might say abstruse – philosophical form invites the extinction of thought itself. Ray Brassier's challenging work, for example *Nihil Unbound* (2007), is probably the best example of this latter, detaching philosophical thinking from the life of the planet to speculate on the possibility of 'thinking a world without thought' (2007: 227; see also Apter 2013: 337). For a more pragmatic view of extinction as a *cultural* as well as a *biological* issue, see Ursula Heise's recent (2016) book *Imagining Extinction*, which argues that contemporary extinction processes, and our capacity to respond to them, are bound up in 'questions of what we value and what stories we tell' (Heise 2016: 5; see also the Preface to this book).

2 Christian theologians and philosophers generally distinguish *eschatology*, the doctrine of last things, from *apocalypticism*, which primarily concerns itself with revelation and uncovering, although both tend to be associated interchangeably in the popular imagination with lurid visions of the 'end times' and/or the 'end of the world'. Apocalypticism, despite the various fundamentalisms with which it is linked, is rife with ambiguity, not least over how to interpret biblical literature, with the Book of Revelation sometimes being read as a means of making sense of worldly suffering, but at other times as a true and detailed prediction of *either* past *or* future events.

3 Akira Lippit's *Electric Animal* (first published in 2000) presents an interesting case for the 'spectrality' of the animal under the conditions of modernity, following Derrida's argument that 'the presence of the animal must first be

extinguished for the human being to appear' (Lippit 2010 [2000]: 8). For Lippit, the 'spectral animal' is at once a symbol and a symptom of modernity. But as Nicole Shukin convincingly counter-argues, animals have not 'disappeared'; rather, they continue – consider the whale – to be substantial material presences and to be exploited as such (Shukin 2009). It is one thing to say that the *idea* of the animal disappears, but another to say that animals *physically* vanish. As Shukin points out, Lippit is more interested in animals as philosophical figures than as physical beings – which risks turning his own argument on its head. See Chapter 4 for further considerations of the 'spectral animal'.

4 The history of whaling in Australia is closely tied up with the colonization of the continent, with British whale ships transporting many of the first settlers, returning later to bring much-needed supplies to early settlements such as the one at Botany Bay (Hoare 2008). As the historical archaeologist Martin Gibbs observes, early whalers and sealers in Australia operated 'on the geographical frontiers of the European expansion into the Asia-Pacific region, as well as on the fringes of the social and economic systems which drove it', helping to establish the conditions for what he calls a 'maritime industrial frontier' (2010: 1).

Such was the case in Western Australia – a periphery within a periphery – where the early history of colonial whaling was played out in almost paradigmatically frontier conditions, and life, by definition precarious, was often nasty, brutish, and short (Gibbs 2010; see also Newton 2013). The boom period was from the late eighteenth to the early nineteenth century, with numerous British, American, and French ships all plying their trade in Australian waters, although by the time the first Western Australian-based companies had come into being in the 1830s, British whaling (as it was elsewhere in Britain's maritime trading empire: see Chapter 2) was already in decline. These companies were short-lived due to a combination of circumstances: fierce competition, especially with American whalers; a lack of capital investment owing to fluctuating oil prices; and depleted whale stocks at the time of market recovery (Gibbs 2010: 18, 117).

The whaling industry, however, would ride out the storm, and by the mid-nineteenth century around 300 whale ships – American, British, French, Australian – were operating off the Western Australian coast and shore stations, with about 40 per cent of the region's total exports being whale products (Newton 2013: 142). The industry would undergo a further lull in the late nineteenth century before the next boom phase in the early twentieth century, when new players (mainly Norwegians) and new industrial techniques would bring in unprecedented numbers of whales: from their Western Australian shore stations, Norwegian crews would take 1125 whales between 1912 and 1916 alone (Newton 2013: 228).

Whaling would continue in Western Australia until 1978, when an 11-metre female sperm whale harpooned by Axel Christiensen – harpooner on the whale-chaser *Cheynes IV* – had 'the sad history of being the last whale to be caught off Albany, the last whale caught in Australia, and the last whale killed by an English-speaking country' (Newton 2013: 260). The *Cheynes IV* survived though, and it now serves as the feature attraction of *Whale World*, a state-of-the-art maritime museum built on the site of the Cheynes Beach Whaling Station, established at Frenchman Bay in 1952 after moving from its original location at Cheyne Beach, and the best-known shore-based whaling station of its time. Winton's novel is attentive to the spirit of this history if not necessarily to its details. It is also well attuned to its many ironies, not least among them the tourist revenue that continues to be generated by Albany's whaling history, which now vastly outstrips the profits originally made from whaling itself (Newton 2013: 268).

5 There is no escaping *Moby-Dick* when writing about whales, and this book is no exception. This has less to do with the text itself, which often goes unread, than with the almost uncanny hold that it continues to exercise over the popular imagination. Philip Hoare, whose previously mentioned 2009 study *Leviathan* is structured around Melville's novel, sees *Moby-Dick* as a 'book made mythic by the whale, as much as it made a myth of the whale in turn. It is the literary mechanism by which we see the whale, the default evocation of anything whalish – from newspaper cartoons and children's books to fish and chip shops and porn stars' (2009: 37). As Hoare points out, *Moby-Dick* – now a magnet for literary critics of all stripes – was not greatly appreciated in its time, and it failed to sell out its first edition (37). Even today, its cult status owes more to the general symbolic importance of the whale than it does to the distinctive myth-making qualities of Melville's writing, which – predictably enough – is often credited with whale-like attributes of its own. Another reason, perhaps, is the hauntingly melancholic effect of a text that resonates across time and place, just as melancholy itself depends on a mythomorphic repetition of basic characteristics that may take up very different forms in different places and at different times. For more on this, see Chapter 4 especially, but see also the various references to *Moby-Dick* threaded through this chapter and Chapter 2.

6 Wry wordplay is a feature of Winton's work. As most of his commentators have pointed out, *Shallows* plays on several different connotations of the title word, which is associated both with the frivolous (shallow thinking) and the hazardous (shallow water) in the text.

7 Winton is especially well known for his advocacy and/or activist work in the area of marine conservation, and he currently serves as Patron of the Australian Marine Conservation Society. Thus, while he continues to support a wide variety of environmentalist causes, many of these are connected to his love of the sea and his more particular desire to protect Australia's underwater heritage against the pressures of human exploitation: rising sea levels, ocean acidification, unchecked coastal development, overfishing – all of which have contributed to increasingly unmanageable levels of marine pollution, habitat destruction, and species loss. At the same time, Winton's writing (and, in some cases, speaking) on these topics is characterized by his lifelong distrust of public institutions and orthodox believers of all kinds. It is therefore noticeable that while he has allowed some of his work to be used by mainstream environmentalist organizations such as WWF, he tends to fight shy of being identified with the mainstream; as he wryly says of his early campaigning days in the 1990s, 'I was, to say the least, an unlikely recruit. To the battle-scarred Birkenstockers of the movement I was a redneck. Actual rednecks on the other hand – friends and neighbours, some of them – thought I'd lost my mind' (Winton 2012: n.p.). Here as elsewhere, Winton seeks refuge in self-deprecation – another feature of his work – but what remains clear is that, for him, an experiential connection with the natural world is the basis for environmental consciousness, and that this connection is deeply *personal*, as it is for his religious faith.

8 This grisly conclusion is withheld from the truncated version of the diaries that finds its way to Cleve Cookson, while Daniel Coupar – the withholder – arguably prevents himself from meeting the same fate (Turner 1993). However, at the end of the novel it remains unclear what has actually happened to Daniel Coupar, and the possibility remains that a history of self-destruction has repeated itself, a pattern throughout the novel that is apparently reinforced by its final image of the stranded whales.

9 As Arne Kalland suggests, contemporary ethical arguments against whaling are complicated by various *cultural* factors – such as those that support the continuation of subsistence whaling in Greenland or the Faroes – but also by those inherently competitive forms of 'symbol-politics' played out by nations that 'want to appear to be civilised states that care about nature' and therefore find themselves duty-bound to speak out against whaling *tout court* (2012: 193). This jockeying for position has taken place on a global stage in which the emphasis has shifted from seeing whales as resources to be exploited to seeing them as individuals to be protected (DeSombre 2007: 147). However, the ongoing history of the International Whaling Commission (IWC) as chief global regulating

body suggests that this transition has been anything but smooth, with universal compliance almost impossible to force, numerous instances of bullying and vote-buying, and an arguably cultural-imperialist attitude to 'weaker' nations and ethnic-minority groups. For trenchant critiques of the IWC, see Burnett (2012) and DeSombre (2007); see also my brief discussions of whaling 'symbol-politics' in the Preface and Chapters 2 and 4.

10 While I have not followed up on this here, it is difficult to resist the temptation to read *Shallows* against the work of the German Jewish philosopher Walter Benjamin, whose famous analogy between allegories and ruins is certainly relevant to Winton's text. Also relevant are Benjamin's equally well-known theses on the philosophy of history: for instance, Angelus and its inhabitants bear more than a passing resemblance to Benjamin's most celebrated figure, 'the Angel of History', who is driven irresistibly into a possibly catastrophic future while piled before him is the wreckage of the past (Benjamin 1969; see also Chapter 4).

11 The 'more-than-human' has become a key concept for contemporary ecological philosophy, confirming the view that humans share the planet with many other species, but also challenging the view that humans have a special dispensation within it. For different applications of the concept, see the work of David Abram (1997) and Donna Haraway (2008).

Works Cited

Abram, David (1997) *The Spell of the Sensuous: Perception and Language in a More-Than-Human World*. New York: Vintage.

Apter, Emily (2013) *Against World Literature: On the Politics of Untranslatability*. New York: Verso.

Benjamin, Walter (1969) *Illuminations: Essays and Reflections*. New York: Schocken.

Brassier, Ray (2007) *Nihil Unbound: Enlightenment and Extinction*. New York: Palgrave.

Burnett, D. Graham (2012) *The Sounding of the Whale: Science and Cetaceans in the Twentieth Century*. Chicago: University of Chicago Press.

Colebrook, Claire (2014a) *Death of the PostHuman: Essays on Extinction, Vol. I*. Ann Arbor, MI: Open Humanities Press.

Colebrook, Claire (2014b) *Death of the PostHuman: Essays on Extinction, Vol. II*. Ann Arbor, MI: Open Humanities Press.

Davis, Mike (1999) *Ecology of Fear: Los Angeles and the Imagination of Disaster*. New York: Vintage.

DeSombre, Elizabeth (ed.) (2007) *The Global Environment and World Politics*. London: Continuum.

Gibbs, Martin (2010) *The Shore Whalers of Western Australia: Historical Archaeology of a Maritime Industrial Frontier*. Sydney: University of Sydney Press.

Haraway, Donna (2008) *When Species Meet*. Minneapolis: University of Minnesota Press.

Heise, Ursula K. (2016) *Imagining Extinction: The Cultural Meanings of Endangered Species*. Chicago: University of Chicago Press.

Hoare, Philip (2008) 'Troubled Waters: Did We Really Save the Whale?' *The Independent*, 19 September www.independent.co.uk/environment/nature/troubled-waters-did-we-really-save-the-whale-935193.html, accessed 25 June 2017.

Hoare, Philip (2009) *Leviathan or, the whale*. London: Fourth Estate.

Hutcheon, Linda (1989) 'Historiographic Metafiction: Parody and the Intertexuality of History'. In P. O'Donnell and R. Con Davis (eds) *Intertextuality and Contemporary American Fiction*. Baltimore: Johns Hopkins University Press, 3–32.

Kalland, Arne (2012) [2009] *Unveiling the Whale: Discourses of Whales and Whaling*. Oxford: Berghahn Books.

Kennedy, Maev (2015) 'Dippy's had a whale of a time – but the party's over'. *The Guardian* 29 January, 22–23.

Kermode, Frank (1967) *The Sense of an Ending: Studies in the Theory of Fiction*. Oxford: Oxford University Press.

Kolbert, Elizabeth (2013) *The Sixth Extinction: An Unnatural History*. New York: Henry Holt and Co.

Lippit, Akira (2010) [2000] *Electric Animal: Toward a Rhetoric of Wildlife*. Minneapolis: University of Minnesota Press.

Miels, Yvonne (1993) 'Singing the Great Creator: The Spiritual in Tim Winton's Novels'. In R. Rossiter and L. Jacobs (eds) *Reading Tim Winton*. Sydney: Angus & Robertson, 29–44.

Morton, Timothy (2013) *Hyperobjects: Philosophy and Ecology after the End of the World*. Minneapolis: University of Minnesota Press.

Newton, John (2013) *A Savage History: Whaling in the Pacific and Southern Oceans*. Sydney: University of New South Wales Press.

Partridge, Christopher (2008) 'The End Is Nigh: Failed Prophecy, Apocalypticism, and the Rationalization of Violence in New Religious Eschatologies'. In J.L.Walls (ed.) *The Oxford Handbook of Eschatology*. New York: Oxford University Press, 191–212.

Pash, Tim (2008) *The Last Whale*. Fremantle, WA: Fremantle Press.

Roman, Joe (2006) *Whale*. London: Reaktion Books.

Schwarz, Hans (2000) *Eschatology*. Cambridge: William B. Eerdmans Publishing Company.

Shukin, Nicole (2009) *Animal Capital: Rendering Life in Biopolitical Times*. Minneapolis: University of Minnesota Press.

Simmonds, Mark P. and Judith D. Hutchinson (eds) (1996) *The Conservation of Whales and Dolphins: Science and Practice*. Chichester: John Wiley & Sons.

Thomas, Roie (2008) *Theistic Existentialism in the Fiction of Tim Winton*. Hobart: University of Tasmania Press.

Turner, John P. (1993) 'Tim Winton's *Shallows* and the End of Whaling in Australia'. *Westerly* 1: 79–85.

Van Dooren, Thom (2014) *Flight Ways: Life and Loss at the Edge of Extinction*. New York: Columbia University Press.

Walls, Jerry L. (ed.) (2008) *The Oxford Handbook of Eschatology*. New York: Oxford University Press.

Winton, Tim (2012) 'Author Tim Winton on his love for the sea and our generation's opportunity to protect it'. http://www.wwf.org.au/74120/Author-Tim-Winton-on-his-love-for-the-sea-and-our-generations-opportunity-to-protect-it, accessed 18 April 2017.

Winton, Tim (1985) *Shallows*. Sydney: Unwin Paperbacks.

2

Sperm Count: The Scoresbys and the North

For Scoresby, [the] greatest significance [of the northern sea] was that it provided whalers with a livelihood that could make an enlightened contribution to the nation.

Michael Bravo ('Geographies of Exploration and Improvement')

[W]e can't read the darkness, Mr Scoresby.

Serafina Pekkala (*Northern Lights*)

Prologue: Double acts

One of the most colourful characters in Philip Pullman's Arctic-oriented trilogy *His Dark Materials* (1995–2000) is the gun-toting Texan balloonist Lee Scoresby, a kind of high northern cowboy from the deep south who, as Pullman suggests, combines the looks and some of the characteristics of the strait-laced nineteenth-century English explorer William Scoresby and the famously sinister-looking American spaghetti-western villain, Lee Van Cleef (www.goodreads.com/characters/1160-lee-scoresby). Pullman's character is no villain, but it would not be entirely appropriate to describe him as a hero either; instead, he exhibits the moral ambivalence – the constitutive doubleness – that is often seen as being one of the principal features of Pullman's text (Lenz and Scott 2015; Rayment-Pickard 2004).

It seems probable that Pullman knows more than he is letting on about the historical Scoresby – or the two historical Scoresbys, to be more precise, for William Scoresby Senior's legendary exploits as a navigator in the Arctic would later be matched (some would say eclipsed) by those of William Scoresby Junior, his son (Bravo 2006; Hoare 2009). And it is probable that Pullman knows more than he cares to admit about the reasons behind the Scoresbys' celebrity. For both father and son, in addition to being explorers of the high north, made their fortunes from the northern whaling industry. And they did so seemingly just at the right time, during the so-called 'heroic age' (Barrow 2001: 2) that would run from roughly 1750 to 1850, even if the younger Scoresby, sensitive to the mounting difficulties caused by a potentially catastrophic combination of increasing competition, increasing risk, and increasingly inadequate technology, was far-sighted enough to retire in the early 1820s, whereupon he became a parish vicar, albeit one – as we will see – with a notably scientific bent. Finally, it seems likely that Pullman has knowledge of this unusual career shift as well, especially given the context of his scathing attack on the established church in *His Dark Materials*, which he also maps onto an equally blistering assault on the follies of Britain's expansionist ambitions in the high north.

In this chapter, I want to look at some of the different discursive strands – especially those relating to commercial whaling, northern exploration, and imperial masculinity – that can be pulled together around one of British maritime history's most intriguing double acts, the Scoresbys. To some degree, my argument follows Philip Hoare's in his excellent 2009 popular historical study, *Leviathan* (see also Preface and Chapter 1). First, like Hoare, I use whaling lore, which I take here as combining scientific and popular understandings, to demonstrate the entanglement of *human* history and *natural* history, with violence (which is as much human on human as human on animal or animal on animal) being at once its most conspicuous by-product and its sometimes emotionally shattering result. Second – and also like Hoare – I place literary representation at the centre of this tangle of historical relations and events, adopting what might be broadly conceived (even if Hoare himself is not especially forthcoming about his own methods) to be an *environmental humanities* approach. This approach, first sketched out in my Preface then developed throughout the book, further challenges the boundaries between *cultural* history and *environmental* history, not least by looking at the reciprocal rather than one-way relationship between humans and animals, one presupposition being that whales have culture, too, and that cetacean cultures and the imaginaries that surround them are neither restricted to nor circumscribed by the human domain.

Finally, I place the Scoresbys – no more than bit players in Hoare's overwhelmingly Melville-inspired account – as key representative figures in the uneasy transition between two broadly defined eras in the history of British whaling: a time when whaling was not yet a full-fledged, fully mechanized global industry, but was no longer (or at least no longer primarily) a regionally based if internationally competitive artisanal trade (Barrow 2001; Bravo 2006; Jackson 1978). William Junior, in particular, was keenly conscious of this transition, and to some extent lived through it. An accomplished, forward-thinking scientist, he recognized the value of developing the innovative techniques that would be needed for whaling to flourish as a mass-scale maritime industry. But at another level, he wanted no part of that emergent industry, which from the early decades of the nineteenth century onwards would not just transform the social and economic conditions of whaling, but increasingly shift its geographical axis of commercial activity from north to south (Burnett 2012; Jackson 1978). This perhaps, in the broadest sense of that always tricky term, might be described as whaling history's 'imperial' moment, even if the Scoresbys' own historically earlier professional ambitions, which acknowledged the political as well as economic advantages of combining whaling trips with (ideally) government-funded journeys of northern exploration and discovery, were bound up whether they liked it or not in a larger, international 'scramble for the Arctic' which had distinctly imperialistic overtones that remain present in British commercial and geopolitical designs upon the Arctic today (Craciun 2009; Huggan and Jensen 2016).

Fathers and sons

The Scoresbys' action-packed lives, like those of many a celebrated father-and-son combination, can be used as a kind of emotional barometer for testing the changing temperature of the times. It is tempting to turn here to one of the most famous British literary examples, namely Edmund Gosse's 1907 memoir *Father and Son: A Study of Two Temperaments*, which was read by its own author as 'the record of a struggle between two temperaments, two consciences and almost two epochs' (Gosse 1983 [1907]: 35), and by some of its later commentators as a modern existential 'manifesto announcing the right of the individual [acting] against the dictates of tradition and parental values to fashion his inner life for himself' (Abbs 1983: 12; for a different view, more sympathetically inclined towards Gosse Senior, see also Thwaite 2002). Still, we should be wary of turning the clock back and using this as a template for the Scoresbys. For while it is certainly the case that William Senior, like Philip Henry Gosse, could be a forbiddingly authoritarian figure, his authority was rarely challenged; and while his son, like Edmund Gosse, undoubtedly had the more subtly enquiring temperament, he never came close to, nor for that matter demonstrated the remotest interest in, shaking the orthodoxies of his inherited Protestant faith. It may also be the case that, like the younger Gosse, William Junior comes across as the 'softer', more impressionable figure, caught between two eras and never entirely comfortable in either. But in some ways, he was very much a man of his time whose university education in Edinburgh would give him access to some of the finest minds of the Scottish Enlightenment (Bravo 2006); and if he remains a quintessentially self-interrogating modern figure, that is less because he consciously breaks with tradition than because he brings together seemingly incompatible traditions, coming to inhabit the uneasy hiatus between the future and the past.

While some attention will be given in what follows to Scoresby Senior (1760–1829), dutifully described by his son as the ultimate practical man, distinguished by his 'extraordinary enterprise' (*Memorials* 121; see also Hoare 2009: 283–287), the spotlight will shine on Scoresby Junior (1789–1857), who was seen no less gushingly by his early biographers as an exemplary man of *action* (navigator and whaler), man of *learning* (scientist and educator), and perhaps above all man of *faith* whose tireless work for the church, arguably outstripping his earlier exploits as a 'learned whaler' (Bravo 2006: 533), spanned a period of over thirty years. Scoresby's nephew Robert Scoresby-Jackson, availing himself of his memoirs three years after his death, would come up with the following comprehensive catalogue of his achievements. Scoresby, he says, was:

> An individual of a singularly active and observant mind, who was ever careful to record and preserve his observations; who in early life enlarged the sphere of his researches by repeated voyages; was the first to make an accurate survey of the east coast of Greenland, and who penetrated further north than any of his contemporaries; who devoted the latter half of his life to the moral instruction and amelioration of his fellow-creatures; a philosopher whose acute intellect embraced some of the subtlest subjects of physical science; a sincere believer and earnest advocate of religious truth, and withal a zealous and indefatigable practical philanthropist. (Scoresby-Jackson 1860: vi)

Hyperbole aside, this is a passingly accurate list that highlights the broad sweep of Scoresby's accomplishments. It also indicates what his nephew would describe as Scoresby's supreme adaptability: equally at home in the rough and tumble of life at sea and in the more refined company of his Wernerian Society scientific colleagues, '[Scoresby] could be the minister, the sailor or the man of science as circumstances required – he could be everything, in short, that was compatible with his Christian profession' (326).[1] Not that Scoresby was as canny as this implies; on the contrary, in his dealings with the Admiralty, in particular, he exhibited a good deal of naivety. This has been interpreted by more sympathetic biographers as a sign of the willingness of senior figures at the time – in particular, the famously manipulative Second Secretary to the British Admiralty, Sir John Barrow – to exploit the goodness of a clever if callow young man for their own personal and professional benefit;

and by less sympathetic ones as a confirmation of Scoresby's egregious lack of political acumen: 'For most of his life', says Fergus Fleming, 'he was to be the establishment's willing dupe' (1998: 32).

Probably the most balanced portrait to date is by the environmental historian Michael Bravo, who positions Scoresby as a Scottish-trained, Protestant-influenced Enlightenment improver whose remarkable scientific achievements in the maritime disciplines of hydrology, oceanography, and nautical magnetism, though critically acclaimed within his own inner circle, were only belatedly given the popular recognition they deserved. Scoresby was an improver in a philanthropic sense as well, and these altruistic sympathies – most evident in later life, but instilled in him at an early age by his father – were rooted in a Protestant work ethic that owed less to his own broad-church Anglican sentiments than to his father's more puritanical strain of Wesleyanism, and which was 'driven by a zealous spirit towards moral and natural truth' that would manifest itself, not just in his religious teachings, but in all aspects of his life (Bravo 2006: 533). Bravo usefully points here to what he calls 'the three profound ontological arenas in Scoresby's world – the oceans, Christian faith, and natural knowledge' (533). He then embeds these in a late eighteenth-century context of 'rapid capitalist expansion' (518) in which the British whaling trade both required and profited from the kinds of 'enlightened, useful knowledge' (519) supplied by an emergent bourgeois class of merchant-navigators like Scoresby, who succeeded impressively in harnessing the traditional skills of the industrial artisan to the modern discoveries of the scientific savant (533).

At the centre of all of this were Scoresby's unshakable religious beliefs, which saw to it that *observation* was dutifully combined with *observance* (as with his father, there could be no question of activity of any kind, commercial or recreational, on the Sabbath), and which had few moral scruples about whaling, the hunting and killing of God's creatures counting as one example among illimitable others of man's divinely sanctioned superiority over the rest of creation within a larger providential context in which all animals, however mighty, would inexorably 'fall victims to the prowess of man and [be] rendered subservient to his convenience in life' (*Account*, vol. 2: 240). Scoresby's unflinching professional dedication to improving the efficiency of methods for catching whales, although it predated the development of some of the techniques that would later become associated with modern industrial whaling, thus needs to be seen in a double context in which *providence* is tied to *progress*, and both of these to the more mundane task of making money, which is as good a way as any of doing God's work. This suggests in turn that, whatever its well-documented flaws,[2] Max Weber's early twentieth-century account of the symbiotic link between the Protestant ethic and the spirit of capitalism (his terms) has multiple resonances with the Scoresbys' work, particularly insofar as it insists on what Weber calls 'the providential interpretation of profit-making, [which in turn] justifies the [day-to-day] activities of the [forward-thinking] business man' (Weber 1974 [1930]: 163). Even closer to the grain is Weber's glancing reference to the acquisitive Wesleyan principles by which good Christians should, in Wesley's apodictic phrase, 'gain all they can and save all they can' (quoted in Weber 1974: 175). Scoresby Senior, an evangelical Wesleyan himself, would no doubt have approved of this; but so too his more accommodating son, who looked forward to what Weber calls the emergence of a full-blown 'bourgeois economic ethic' (177):

> With his consciousness of standing in the fullness of God's grace and being visibly blessed by Him, the bourgeois business man, as long as he remained within the bounds of formal correctness, as long as his moral conduct was spotless and the use to which he put this wealth was not objectionable, could follow his pecuniary interests as he would and feel that he was fulfilling a duty in doing so. (176–177)

As is well known, Weber's work involves an attempt to sketch out the lineaments of a particular, Western form of capitalism that is backed up by the latest developments in the natural sciences and 'the rational ethics of ascetic Protestantism' (27). There can be no doubt that these two support mechanisms were integral to the professional careers as well as personal world views of the Scoresbys, especially William Junior, and if it might be pushing things a little to describe either man as 'ascetic', in neither case was there a contradiction (and Weber certainly wouldn't have seen one either) between the strictures of rational asceticism – temperance for instance – and what Weber calls the 'impulsive enjoyment of life' (167).

A further connection exists between the Scoresbys and two of Weber's representative rationalist types, the 'colonial entrepreneur' and the 'capitalistic adventurer' (20), although the inevitable sketchiness of both types fails to allow for the kind of detail needed to distinguish William Senior from William Junior, or to account for the complexities of either figure. Notwithstanding, in the case of both Scoresbys, the adjective 'practical' was routinely invoked to cover many of the activities they pursued as well as to justify the violence, explicit or implicit, bound up in them. Hence Scoresby Junior's extravagant praise for his 'practical' father, who is presented in his 1851 biographical account as a paragon of common sense if not a man (like himself) of social refinement or formal learning; as a scientist by default rather than by design whose formidable 'applied intelligence' helped turn him into one of the most successful whalers of his times (*Memorials* 213). And hence the equally lavish praise that has often been meted out to Scoresby Junior himself, who was scarcely less successful as a northern whaling captain than his father, and more successful still in working out 'how to make whaling a viable vehicle for the scientific exploration of the Arctic' (Bravo 2006: 533): a parallel achievement that has since arguably earned him his place in the pantheon of famous British navigator-explorers even if, repeatedly marginalized as he was by the Admiralty, this recognition largely escaped him during his life.

Not a violent man by nature, Scoresby Junior is seen by most of his biographers as being somewhat timid and reserved, even if some of his more audacious exploits at sea gainsay this; however, he was directly involved in a business that was as violent as any in his day. In true hagiographic style, Tom and Cordelia Stamp try to downplay this by calling him a 'friend rather than an exploiter of nature' (1976: xii), but this depends on a rather charitable view of whaling that Scoresby himself would have been highly unlikely to endorse. On the contrary, in his scientific books, notably his two-volume *Account of the Arctic Regions* (1820) – the most comprehensive study of its time – Scoresby is only too clear about the dangers involved in hunting whales, speaking with some sensitivity about the pain involved for both whalers and whales themselves in capturing and destroying the great animals, only to insist almost immediately that 'the object of the adventure [and] the value of the prize' are too great 'to be sacrificed to feelings of compassion' (*Account*, vol. 1: 472). Here as elsewhere in his substantial body of written work, Scoresby is supremely aware of the violence of the whaling enterprise and its risks for humans and animals alike, but he also talks counter-intuitively about the 'joy of capture' (*Account*, vol. 1: 472). In the next section of this chapter, I want to associate this 'joy' with the spectacle of whaling – a history, both experiential and representational, in which violence has long functioned as a standard commodity – and with what Scoresby calls the 'beautifulness of manly excitement' bound up in the seafaring life (quoted in Scoresby-Jackson 1860: 378).

Beauty and the beasts

Scoresby begins the second volume of his Arctic *Account* with a thumbnail sketch of Western whaling history. Framed within the wider context of a providential narrative of humanity's epic encounters with fearsome 'beasts of prey', this history is traced in terms of humanity's progress from prey to predator, with an early 'waging [of] a war of death' with the whales eventually giving way to a gradual improvement in the industrial techniques needed for their harvest, and with the ancient enemy eventually being transformed into a profitable commodity via a grand narrative of overcoming that is seen as integral to the establishment of a global maritime system of mercantile trade (*Account*, vol. 2: 1–3; see also Bravo 2006). The history of whaling, for Scoresby, is enterprise in its broadest sense, born of both adventuring spirit and business savvy, while it is also incorporated into the parallel histories of *mercantilism* (the pursuit of profit in the national interest) and *imperialism* (the frequently violent playing out of national rivalries in which commerce is contingent on discovery, and both of these depend on creating favourable, that is competitively advantageous, conditions for trade).[3]

Key to all this is Scoresby's understanding of *providential design*, which he doctrinally interprets in terms of a 'power given [to man] over created nature' that is directly attributable to the circumstances of Creation itself (*Account*, vol. 2: 241). In accordance with this pre-Darwinian understanding of nature's bounty – and nature as bounty – each successful killing of a whale is adjudged to be a 'joyful circumstance' (*Account*, vol. 2: 249). Not that this joy is necessarily apparent in Scoresby's dispassionate account of the modern killing and flensing process, which is shorn of the drama of his earlier, racier accounts of primordial encounters between human beings and whales. Instead, the latter are now to be seen for what they are (objects of commerce) rather than what centuries of 'superstition and dread' have made of them (monsters of the deep). Once caught and cut up, they are little more than hunks of bone and blubber, with the drama consisting rather in getting the product back to port as expeditiously as possible so that the profits can be secured without sustaining damage to equipment or loss of life (*Account*, vol. 1: 451).

This is not to say that Scoresby doesn't find whales to be beautiful. On the contrary, as both ancient living beings and modern scientific specimens they are objects of wonder. This wonder then only magnifies as the animals' physiological and biological complexity, which Scoresby was astute enough to see had been drastically under-appreciated, is increasingly understood (*Account*, vol. 1: 446–447). In the first volume of the *Account*, in particular, Scoresby proves himself to be an adept natural historian; and while he might not have been especially competent at drawing whales, he was typically efficient at describing them, showing a natural historian's eye for physical detail together with a natural historian's sensitivity to the minute as well as the massive that later cetacean appreciators, Melville almost inevitably among them, would see as being characteristic of his work (Morgan 2016: 8). There is a beauty in facts that Scoresby's contemporaries would have immediately understood, and that more recent commentators have belatedly appreciated: the 'aesthetics of data', the literary critic Benjamin Morgan winningly calls this, finding a fitting example in the distinctively inter-animating combination of '[intimate] quantitative description and [codified] aesthetic experience' to be found in Scoresby's popular scientific work (2016: 11; 16).

There is beauty of another kind, however, in Scoresby's oeuvre, which pertains to the codified realm of *male* experience. It is no surprise, of course, to find whaling being used to celebrate a certain, stereotypical kind of masculine endeavour (toughness, single-mindedness, bravery, etc.), although much of the classic whaling literature – again often taking its cue from Melville – has tended as much to undermine as underscore this, implicitly rejecting 'the masculine mystique by [explicitly] dramatizing [...] male heroism's awful result' (Baurecht 1986: 53; for a fictional example that draws extensively on the Scoresbys, see also Gaskell 1996). If, as Caleb Crain has it, sperm rather than sperm oil is the natural currency of many a whaling narrative, then Melville's persistent jokes about it tend to run closer to tragedy than comedy, eventually turning *Moby-Dick* into a multipurpose allegory for the violence that materializes when industrial capitalism drains any 'higher spirit [that might be] found in [platonic] moments of community between men' (2012: 11). We shouldn't expect to find emasculating accounts of this kind in Scoresby's work – and we do not – but in occasional passages like the following one we can sense a threat to the very ideals of masculinity and male camaraderie that are being presented to us. Here, during the description of a storm (taken verbatim by Scoresby-Jackson from Scoresby's memoirs), Scoresby picks out four men tending to the wheel of a listing ship, the *Royal Charter*:

> Every man there [Scoresby says] was the picture of energetic manly life. You saw in his face an expression, to be read of every one, that he felt that in the management of the wheel he held the destinies, under Providence, of ship and human life in his hands – and that the ship broaching to by his failure in seamanlike tact and management would be destruction; hence the picturesque adaptation of his strength and figure showed, in nature's true developments of expression and *beautifulness of manly excitement*, that the importance of his performance and duty was fully realized. (Scoresby, quoted in Scoresby-Jackson 378, emphasis mine)

On the face of it, this scene represents probably the clearest celebration in Scoresby's work of the sailor's life as an *aesthetic* as well as a *moral* enterprise, an illustrative example of the ways in which – and the appeal with which – a group of hard-bitten men, brought together under God's guidance, are subjected to gruelling trials that are at one and the same time providence's chosen mechanism for testing their capacity for collective physical effort as well as their individual religious faith. But it is also clear that providence holds the cards, and that for all Scoresby's championing of homosocial endeavour (which anticipates Conrad's 'fellowship of the craft' rather than Melville's more contemporaneous, also more sexually ambiguous, seafaring community),[4] such efforts are strictly subordinate to his belief in the controlling will of God.

Muscular Christianity might be a less ambivalent way of describing this, if without the larger (British) imperial context that would later help define the twin Victorian myth models of the 'Imperial man' and the 'muscular Christian', a dyad usually seen in terms of an amalgam of 'manly' qualities: 'physical prowess', 'intellectual energy', 'moral purpose', and all in the service of national pride (Benyon 2002: 27; see also Hill 2008). Some caveats are needed here. For one, it would not be strictly accurate to describe either Scoresby as an imperialist. However, they were both staunch nationalists throughout their lives and their various maritime achievements, especially in the field of northern exploration, were shamelessly exploited by the Admiralty as an exercise in imperial public relations, as for that matter was the theatre of British Arctic operations at large (Fleming 1998; Hill 2008). And for another, muscular Christianity as a particular, identifiably late Victorian form of British imperial masculinity doesn't quite fit the Scoresby mould – and not just for the obvious reason that they both significantly predated it, but also because, for them, imperial narratives of ambition were considered secondary to providential narratives of improvement that saw God's work on earth as a shared human duty, albeit one that required strong leadership by men of unsullied virtue, enlightened knowledge, and experience-backed common sense.

Still, there is a degree to which Scoresby Junior, in particular, looks beyond his time, whether by combining Skoumand's four principal aspects of the stereotypical Victorian gentleman figure, namely the 'officer and gentleman' (Scoresby as captain), the 'scholar gentleman' (Scoresby as scientist), the 'Christian gentleman' (Scoresby as cleric), and the 'gentleman-sportsman' (Scoresby as explorer), or by anticipating the supposedly milder characteristics that would later come to popularize the equally typecast figure of the 'new man' (Skoumand, quoted in Benyon 2002: 29–30).[5] Whatever the case, it needs to be borne in mind that neither Scoresby was a progressive in most accepted modern senses of the word. Quite the inverse, both men were deeply conservative in outlook, shaped as they were by their strict religious upbringing as well as by a lineage of Scoresby patriarchs who prided themselves on their competitive combination of physical strength, intellectual enterprise, and moral restraint (*Memorials* 22).

There was a strong element of *regional* pride as well, with both Scoresbys finely attuned to the northern locations of their birth (Whitby) and education (Edinburgh), as well as compulsively attracted to the northern latitudes (the Greenland Seas) that would provide the primary theatre of exercise for their life's work. As I will now go on to discuss, the Scoresbys also contained, deeply embedded within both their personalities and their philosophies, some of the ideas and ideologies that have come over time to be popularly associated with the north. As the literary scholar and art historian Peter Davidson has argued, there is no such thing as a single idea of north. Rather, there are multiple, conflicting ideas and imaginings that have proved hugely influential in shaping attitudes and perceptions which, in turn, have driven activities in the Arctic and other high northern regions that have made vast fortunes for some and early graves for others, providing a rich site for ambitions both achieved and not, adventures both brilliantly realized and spectacularly backfired (Davidson 2005; see also Craciun 2009).

Ideas of north

Near the beginning of the first volume of his 1820 *Account*, Scoresby Junior places himself in a long line of British Arctic discoverer-explorers (Davis, Baffin, Hudson), praising their 'laborious exertions, notwithstanding the many disadvantages under which they were conducted' (*Account*, vol. 1: 22). This, in turn, gives us a good sense of his own lifelong commitment to the practical prosecution of intellectual curiosity and hard work. 'Curiosity' is euphemistic, perhaps, given Scoresby's well-trained eye, inherited from his father, for the latest commercial opportunity, though true to form he sees a larger, providential design behind northern voyages of discovery and exploration, which may serve mercantile and/or geopolitical interests but are still primarily to be seen as ways of extending human knowledge for the morally upstanding purpose of better performing God's work on earth (*Account*, vol. 1: 23). Again like his father, Scoresby Junior was perfectly well aware of the risks involved in pushing ever further north, whether in search of the North Pole or the North-West Passage.[6] Both of these grail-like goals he saw as being accessible, if only under the right circumstances, but he also recognized the need for men such as himself, who were practically minded and professionally experienced, to apply their growing scientific understanding of the region to the perilous task of navigating Arctic waters, which required 'an extensive knowledge of the nature, properties, and usual motions of [sea] ice' (28).

Not that the Admiralty was listening, and while both Scoresbys would lead a number of successful expeditions to the high north, sensibly offsetting their costs by combining them with whaling and hunting (Bravo 2006: 533), the younger Scoresby was repeatedly snubbed in his plans, either by being seen as a professional whaler rather than as a suitably qualified naval commander or (what amounted to the same thing) by being overlooked in favour of a mostly dismal procession of establishment-backed figures with precious little practical experience of the Arctic and still less common sense (Fleming 1998; Stamp and Stamp 1976). Fergus Fleming is especially harsh in his judgement of the man who was almost invariably at the centre of these decisions, Sir John Barrow, whose unrivalled enthusiasm for Arctic exploration was only matched by his singular incompetence: 'Perhaps no man in the history of exploration', Fleming witheringly remarks of Barrow's obsession with the North-West Passage, 'has expended so much money and so many lives in pursuit of so desperately pointless a dream' (1998: 423).

Fleming probably goes too far in seeing the history of British Arctic exploration *tout court* as an unparalleled disaster – a toxic combination of daredevil adventure, political intrigue, and out-and-out farce, with the North-West Passage headlining as 'an eternal symbol of futility' (419) – but some of his general points still hold: that far too many expeditions achieved far too little; that many of them were spectacularly mismanaged; and that the imperial fantasies of discovery and conquest that sustained them were almost by definition unfounded, with loss of life, equipment, and money – hastily repackaged as 'heroic failure' – being one of the most notable results (423; see also Hill 2008). Significantly, Fleming grants a rare exemption to Scoresby Junior (listed among his pantheon of British Arctic explorers at the end) on the grounds that while he was an 'Admiralty pariah' (444), he was also a man of considerable talents whose practical experience and specialist knowledge would have greatly benefited the very establishment that chose so foolishly to ignore him, stubbornly refusing to see him as one of their own. But this again appears to be overstating the case; for as Michael Bravo suggests, Scoresby in fact 'played multiple and significant roles in Arctic exploration during the Regency' (2006: 533), advising such scientific luminaries as the naturalist Sir Joseph Banks, helping to reform the government regulations surrounding financial rewards for exploration achievements, and offering a viable funding model for Arctic expeditions that was duly taken up by several of his contemporaries, although, marginalized as he continued to be by the Admiralty, it would take him the best part of two decades to set the record straight (533).

Fleming and Bravo are agreed, though, that Scoresby's reputation today has been partially restored, even if when set against the semi-mythical likes of Ross and Franklin he cuts a minor figure, still better known for his whaling exploits than his scientific discoveries, and still relatively neglected in the history of British Arctic exploration despite his startling achievements in (to cite just one of many possible examples) mapping some of the more intractable stretches of the jagged East Greenland coastline, an extensive area of which now bears his name: 'Scoresby Sound'. Scoresby deserves credit, too, for his *written* work, which, while of uneven literary quality, still succeeds in shedding insight into the Arctic imaginary of his early nineteenth-century contemporaries as well as the various, often politically motivated ideas of north that have been invested for many centuries in the Arctic, which remains 'as much ideological as physical terrain' (Hill 2008: 3; see also Ryall, Schimanski, and Wærp 2010). As Peter Davidson argues, it is the *doubleness* of north that has helped attract generations of literary visitors to the Arctic, most of whom have had little intention of actually going there (2005: 46–47). This ideational north, he suggests, comprises a vast, internally differentiated realm that includes both all-too-real and preposterously imagined places. It is also a highly dangerous space in which the mental traveller, confronted with the outer limits of his or her own imaginative capacities, runs the risk of over-reaching them, stepping into potentially life-threatening territory where 'beyond the fields we know, there may either be horrors or islands of the blest' (25; see also Huggan 2016: 335–336).[7] Such doubleness is a defining feature of the travel narratives that would have been instantly familiar to many of Scoresby's contemporaries, and which helped to turn the Arctic into something of a public obsession during the first half of the nineteenth century (Garrison 2008: 381). This obsession was later consolidated in the Victorian period, when it fed into imperial narratives of heroic adventure sustained by a burgeoning public interest in what Jessica Richard calls 'the business of empire', which was a natural consequence of the competitive expansion of de-monopolized regional trade (2003: 297).

How do we fit Scoresby Junior's work into this imperial romantic context? One possible answer to this is that we don't need to. After all, Scoresby himself would have had little patience with Arctic/northern imaginaries of the kind sketched above, just as he had no time for popular explorers' accounts, the standard embellishments of which are sternly brushed aside in the first volume of the *Account* as reprehensibly self-serving instances of a 'prevailing desire [...] to communicate extraordinary circumstances [which] has a tendency to bias the judgment of the most candid person' (*Account*, vol. 1: 44). As should be clear by now, Scoresby hated showing off, and much of the *Account* is written expressly against this; instead, it sets out its stall by the quality, quantity, and above all *accuracy* of the multidisciplinary information – historical, geographical, biological – it supplies. As Robert Morgan contends, Scoresby's two-volume text, which was widely read and admired at the time, is best understood as belonging to the popular genre of the nineteenth-century descriptive account, which is a literary vehicle for the presentation of factual information; thus, while it plays to some extent on its readers' expectations of a maritime voyage narrative, it is first and foremost a fact-based compendium that should be read – and demands to be read – 'for the scientific data [it] contain[s]' (2016:1). Expansive, general, informed: sweeping natural-historical accounts of this kind were generally aimed at an educated if non-specialist audience that they then sought to persuade of the calculated rationality of their arguments by alternating between appropriately unostentatious forms of storytelling (Scoresby's *Account* tends almost automatically to apologize for its periodic lapses into sublime or dramatic registers) and the patient explanation of biological and/or geophysical phenomena via technical language that was specialized enough to carry scientific authority, but not so specialized as to alienate or talk down. Needless to say, there is nothing transparent or literal about this; and nothing 'non-ideological', either, insofar as the dominant interpretative framework – natural history – comes equipped with what Mary Louise Pratt calls a 'benign imperialism' (1992: 34) that expands the supposedly innocent art of scientific classification into the kind of systematizing fervour by which 'every species on the planet [is extracted] from its particular, arbitrary surroundings and [is then placed] in its appropriate spot in the system (the book, collection, or garden) with its new, secular European name' (31).

But what kind of Arctic is Scoresby's? And what kind of north? As might be expected, there is no easy answer to this: for the Arctic of the *Account* is both richly imagined and vividly real, both geographical *oddity* (more strange than sublime) and scientific *opportunity* (for such strangeness to be explained). Scoresby's Arctic – and, more generally, his north – is above all a working space, resembling at times nothing less than a scientific laboratory in which different biological, hydrological, and meteorological experiments are to be conducted, both for the general furtherance of human knowledge and the particular reconfirmation of God's plan. At one level, Scoresby's experiments are vigorous empirical pursuits of first cause – recorded observations of how things come to be, and of what effects are produced by the action of one thing upon another – but at another, they are humble acknowledgements of a divine sanction that supersedes the 'accuracy [that even the best of scientific] opinion, experience and observation [can] confirm' (*Account*, vol. 1: 397). As Alison Winter observes, 'for Scoresby, the study of nature [is] a heroic expression of one's religious calling' (1998: 254, also quoted in Morgan 2016: 7), while Benjamin Morgan draws attention to the 'compatibility of religion and science' in his work (2016: 7). Both are right, although neither of them mentions the contemporary word that seems best to describe this dual impulse: the American biologist E.O. Wilson's term 'consilience', which he uses to call for a much-needed reconciliation of 'scientific' and 'religious' apprehensions of the phenomenal world (Wilson 1998). Then again, Scoresby would have seen no need for reconciliation at all since, for him, the ultimate authority of God is unchallenged. What remains for His servants on earth is to understand His workings and, in so doing, to admire the profusion, intricacy, and infinite variety of what He has created – and to admire it all the more for its reaching out even into 'the regions the most remote from human observation', thereby rendering the Arctic as simultaneously an *exceptional* and an *exemplary* domain (*Account*, vol. 1: 477; see also Morgan 2016: 7).

This Edenic vision of plenitude – a version of the sublime after all – extends to the animal kingdom as well, as would later be captured in Oliver Goldsmith's absurdly overcrowded illustrations (a pictorial effect of natural-historical inclinations towards the encyclopaedic, styled self-consciously after Scoresby) of Arctic marine life (Goldsmith 1870; see also Hoare 2009: 266). In fact, for all Scoresby's fascination with whales, there is a fully engaged attention to all creatures great and small throughout his writings. In one typical sequence in the first volume of the *Account*, he relishes the opportunity to perform experiments on the body of a tiny 'medusa' (jellyfish); to calculate composite measurements on the basis of these experiments; then, extrapolating from all this, to praise the 'immensity of creation, and the bounty of Divine Providence, in furnishing such a profusion of life in a region so remote from the habitations of men' (*Account*, vol. 1: 179).

Here as elsewhere, Scoresby's (scientific) discoveries are not necessarily new, nor are the (religious) conclusions he draws from them; what is remarkable, if still conventional enough for the particular descriptive genre he is drawing upon, is the range of topics covered and the breadth of knowledge contained. The *Account*, in this and other respects, has both the advantages and disadvantages of a miscellany; and it articulates a similarly miscellaneous view towards a natural world that, in scriptural fashion, it reads in terms of a complex tracery of conventional signs. These signs all point (as they must) to God, but they also have other commonly accepted referents: hence Scoresby's occasional recourse, particularly in geophysical descriptions, to contemporaneous aesthetic conventions of the sublime and the picturesque. Here, for example, is an early passage from his description of the island of Spitzbergen (modern-day spelling Spitsbergen), which seems to come straight out of the manual of eighteenth-century environmental aesthetics:

> This country exhibits many interesting views, with numerous examples of the sublime. Its stupendous hills rising by steep acclivities from the very margin of the ocean to an immense height; its surface, contrasting the native protruding dark-coloured rocks, with the burden of purest snow and magnificent ices, altogether constitute an extraordinary and beautiful picture. The whole of the Western coast is mountainous and picturesque; and though it is shone upon by a four months' sun every year, its snowy covering is never wholly dissolved, nor are its icy monuments of the dominion of frost ever removed. The valleys opening towards the coast, and terminating in the back ground with a transverse chain of mountains, are chiefly filled with everlasting ice. (*Account*, vol. 1: 94)

Such scenic views are disrupted, however, when Scoresby proceeds to recall one of the numerous opportunities he has had to step on shore. Again, the conventional signs are all present and correct, as Scoresby climbs high to a position from which, 'as if elevated to the heavens', he can contemplate being 'superior to the dangers to which I was surrounded' – a textbook definition of the sublime (128). But he and other members of his landing party are soon in trouble once the descent is commenced: now 'the way seemed precipitous' (129), and the men, forced to cross a stretch of solid ice, are struck with 'astonishment and fear' (130). This about as dramatic as the *Account* gets, at least in Volume 1, but what is most revealing about it is Scoresby's tacit recognition of the limits of his own descriptive vocabulary; it is at such moments, as Morgan puts it, that Scoresby turns away from the 'controlling poetics of the sublime and the heroic' to articulate a 'more ambivalent aesthetic' that captures the strange and unexpected, turning these into sources of anxiety and self-doubt (2016: 8, 10). Morgan is probably labouring the point insofar as such ambivalence is of a piece with standard literary accounts of the high north (Davidson 2005; Huggan 2016), but he is still right to point out the difficulties this poses for Scoresby, whose explanatory impulse – a function of the calculated rationality that governs his Arctic expeditions – is short-circuited upon confrontation with the seemingly inexplicable or unknown.

That this same rationality underpinned Scoresby's experiences as a whaler should be clear enough by now, as should his understanding of the symbiotic relationship between Arctic scientific research and commercial whaling: for the implementation of one improved the effectiveness of the other, also helping to create more favourable conditions for the safe exercise of both. Such research, Scoresby firmly believed, had practical implications for improving the often desperate living and working conditions that were the daily reality of those who turned professionally to Arctic whaling, which by definition was an extremely hazardous pursuit in a no less dangerous part of the world (*Account*, vol. 1: 444). Accidents on the job were rife, fatalities were common, and there was always the risk of cataclysms, such as sudden shipwreck or, still worse, protracted tragedies such as those of 1835–1836 and 1836–1837, when a number of whale ships were forced to winter over in the Davis Straits (off the west coast of Greenland) and, lacking sufficient food and clothing, many of their crew members expired (Barrow 2001: 89).

By then, Scoresby had long retired from professional whaling, which – in the Greenland fishery at least – was in its own death throes as a combined result of 'free trade, commodity substitutions and increasing difficulties encountered by the whalers themselves' (Barrow 2001: 93). To what extent Scoresby foresaw this demise is a matter of dispute among maritime historians: Tony Barrow, for instance, sees him as having been aware of the threat but not necessarily attentive to it, although the timing of his retirement might suggest otherwise (2001: 93; see also Bravo 2006 and Jackson 1978). Whether Scoresby was exercised by the broader *ethical* implications of whaling seems less contentious – he almost certainly was not – although his recorded experiences as a whaling captain (anticipating his later philanthropic work for the church) suggest that he had a lively concern for the safety and welfare of his men. The safety and welfare of the whales themselves was another matter, although, practical businessman that he was, Scoresby was both alert to and informative about the dangers of overfishing certain areas, which had the effect of driving the whales further northwards – into increasingly precarious territory – as well as potentially reducing stock to levels that were not only no longer profitable but no longer capable of being sustained. The rights and wrongs of whaling itself were highly unlikely to have troubled Scoresby's sleep; and while this was the norm for whalers of his times, it is still worth enquiring into the discrepancy between his acute moral intelligence and his seemingly chronic lack of understanding for an industry that, for most contemporary observers, has always borne the seeds of its own destruction, eventually ushering in its own extinction even as it came to threaten the existence of the very animal it depended upon: the whale (Burnett 2012; Hoare 2009; see also Chapter 1).

Greenland rights and Greenland wrongs

Whaling was a 'filthy business' (Hoare 2009: 124). Relentlessly profit-driven, it sought to turn oil into gold, transforming the professional whaler into what Philip Hoare memorably describes as a kind of 'pirate-miner – an excavator of oceanic oil, stoking the furnace of the Industrial Revolution as much as any man digging coal out of the earth' (108). It is perhaps above all the sperm whale that Hoare has in mind when, towards the end of his wide-ranging historical study, he numbly contemplates the variety of ways in which 'the cetacean population of the world [has] been hunted, harpooned, blown up, ground down and consumed in a manner unrivalled by any other exploitation of the earth's living resources' (349). For, as Hoare suggests, sperm whales are the most iconic cetacean species of all: 'ask a child to draw a whale, and he [*sic*] will trace out a sperm whale, riding high on the sea' (65); ask an adult, and he or she (but still most likely he) may well reference Moby-Dick, whether he or she has read the book or not. Sperm whales are not only the most ancient of whales but also the most imagined. Animals whose lives 'came to be written only because [they were] taken', they are 'so wreathed in superlatives and impossibilities that if no one had ever seen [them], we would hardly believe that [they] existed' (65). No wonder, Hoare wryly remarks, that Melville's novel, a mythological tale about a monomaniac, would come to haunt later writers and still continues to exercise a massive popular influence today (66).

Sperm whales, however, would have little effect, real or imagined, on the Scoresbys. After all, when *Moby-Dick* was first published in 1851, Scoresby Senior was long gone, and Scoresby Junior would only have a few years left of his life. (Both Scoresbys were well known to Melville, though, who praised the latter and poked fun at the former, giving him a suitably ridiculous cameo role in the novel as the storm-tossed veteran Greenland whaler, Captain Sleet.) More to the point, sperm whales were neither man's main target in the Greenland Seas, where they were few and far between; instead, the principal quarry was *Balaenus mysticetus*, which today would be known as the bowhead, but in the Scoresbys' time usually went by the name of the common whale or, in its more technical-sounding if no more scientifically accurate version, the Greenland right.[8] Greenland rights, like other right whales in different parts of the North Atlantic, were so known because they were the best ones to capture. As the economic historian Gordon Jackson asserts, Greenland rights were reasonably common, suitably plump, and relatively easy to find; they were also sluggish movers, with 'all the docility of a grazing animal and none of its timidity', plus the added advantage of being buoyant once caught and slaughtered, a major occupational hazard should – as was the case with most other cetacean species – the situation be the reverse (Jackson 1978: 8).

This basket of attributes was not lost on the Scoresbys, who gave thanks to God for the 'great First Cause' that had originally granted man power over nature (*Account*, vol. 1: 241). God's grace was also made manifest, William Junior wrote, in the relative tameness of some of the earth's largest inhabitants, which duly allowed them to succumb to human ingenuity and courage, although this last asset, he was generous enough to acknowledge, was evenly demonstrated on human and animal sides (*Account*, vol. 2: 472). Irony comes too easily, perhaps, and Scoresby Junior deserves considerable credit for his zoological work on whales, which was assiduous and (mostly) accurate without being unduly authoritative: like the good scientist he was, he was always aware of what and how much he did not know (*Account*, vol. 1: 446). He was ahead of his time as well in recognizing that certain, observable aspects of cetacean behaviour, for example the ability to remember and adjust, or the affection shown by whales for their offspring, were not just worthy of respect, but 'would do honour to the superior intelligence of human beings [themselves]' (*Account*, vol. 1: 472).

This unashamedly anthropocentric view of the natural world might seem odd today at a time when it is generally, if by no means universally, accepted that non-human animals have culture, and when calls are increasing for their moral standing to be accounted for, with some, including cetaceans, now being considered entitled to the personhood that human beings have historically reserved for themselves (Whitehead and Rendell 2015: 300; see also Chapter 4). Still, the occasional glimmerings of sympathetic imagination shown in Scoresby Junior's work were fairly unusual for his time, and were almost unheard of for a whaler. This again raises the question of his view of whaling. A few facts and figures may help put this into perspective. As Tony Barrow's research on the north-east England whaling trade has shown, the Greenland fishery was a massively lucrative source of whale oil, supplying nearly half of the oil sold on the British market by the mid-1780s (Barrow 2001: 35). Most of this was supplied by northern English ports, with the number of Greenland-bound whale ships increasing fivefold between 1783 and 1787 alone (35). Whitby, the port from which the Scoresbys operated, accounted for a good portion of the Greenland trade, which increased exponentially during the first two decades of the nineteenth century, with the Scoresby-skippered vessel, *Resolution*, being by far the most successful in the fleet (53). Indeed, Whitby in general and the Scoresbys more particularly were nothing if not Greenland specialists, with the entire Whitby fleet sailing to the Greenland Seas in 1819 and 1820 (93).

Given the gains to be made, the place they grew up, and the values instilled in them by their upbringing, it was not surprising that the Scoresbys should turn their hand to whaling, with both father and son counting among the most successful whalers of their times. Scoresby Senior, in particular, was a major talent, capturing a record thirty-six whales on the first voyage for which he acted as captain, and going on during a half-century career to 'claim the lives of no fewer than 533 whales' (Hoare 2009: 283). The father's values were transmitted to the son, including the principle of 'Divine enactment', which decreed that the catching and killing of whales was not just a highly profitable enterprise, but also a righteous pursuit (Hoare 2009: 286). In many ways, William Junior was destined to join his father, having sailed on the *Resolution* from the age of fourteen (Hoare 2009: 286; see also Scoresby-Jackson and Stamp and Stamp). However, as previously explained, the Scoresbys were by no means one and the same, and biographers have sometimes made too little of the tensions that existed between them. It is true that, in many respects, William Junior could be seen as 'a refined version of his father: rational, scientific, pious, inquisitive; a combination of disciplines that the era allowed' (Hoare 2009: 287). But at the same time, William Junior, as I hope to have shown, was both an externally multifaceted and an internally conflicted figure; and though his well-documented sensitivities did not always go down well with his father, they still give us insight into some of the issues that were exercising many of his scientific colleagues at the time.

One burning issue was how to improve the efficiency of industry by putting scientific discoveries to the test – and, as might be expected, Scoresby Junior was exceedingly interested in the effect these latest discoveries might have on whaling, and on oceanic navigation as a whole. (In his own specialist area, nautical magnetism, he would help improve the navigational accuracy that was intrinsic to the success of northern whaling and northern exploration alike.) However, he was also keenly aware that times were changing, and not for the better as far as the northern whaling trade was concerned. The early nineteenth century saw a general downturn in northern whaling – although there were still some spectacular individual successes – with the withdrawal of government support, the spread of coal-gas lighting, and some heavy losses in appalling weather conditions all contributing to the widespread perception of a regional trade in possibly terminal decline (Barrow 2001; Jackson 1978). Scoresby had these misgivings, too, and he was conscious that Greenland whaling was as dangerous as it had ever been, but no longer as profitable as it once was. It remains unclear to what extent his retirement was timed, but it certainly turned out to be prescient, and although the Greenland trade would continue well into the 1830s, the birth of modern industrial whaling – which would look increasingly to other parts of the world to turn its profits – saw to it that the kind of life that he and his father had once taken for granted was all but dead (Jackson 1978). What Scoresby emphatically did not foresee was the devastating impact that whaling would come to have on whales, though unlike his father he seems to have been aware at some level that nature's resources were not unlimited, and that the sustainability of cetacean species (even of the *right* species) could not be vouchsafed. That awareness was not widely shared, with far-reaching consequences for right whales, whose numbers dwindled alarmingly in some places; and although the North Atlantic right would go on to become the first cetacean to be officially protected in the early twentieth century, contemporary marine biologists are increasingly concerned that its 'gene pool is now so restricted that it is unlikely to survive' (Hoare 2009: 207).

Epilogue: Scoresby sites and Scoresby Sound

Lest the Scoresbys be forgotten today, it might be recalled that they were once minor celebrities, and that neither of them was exactly averse to reminding the world of the significance of his own life. That significance has been underscored, in turn, in sycophantic biographies by relatives (Robert Scoresby-Jackson); celebratory accounts by local historians (Tony Barrow; Tom and Cordelia Stamp); and in the various, more or less self-serving memoirs and memorials bequeathed, not least to each other, by the Scoresbys themselves. The Scoresby legacy is also visible in other forms: in the twin portraits in Hull's Maritime Museum; in the now mostly dismantled whale-bone buildings and arches of Whitby; and in the Gothic monument erected to William Junior in Torquay (his resting place), in the church where he had latterly served as an Honorary Lecturer, which features 'an open Bible, a ship and anchor, a mariner's compass, and other mathematical instruments' in a panel on its base (Scoresby-Jackson 1860: 399). The symbolism is appropriate for a man who, not always fully appreciated in his day, now seems 'assured of a place in the history of science [for] his pioneering work in the Arctic and [...] the science of the sea' (Stamp and Stamp 1976: 246; see also Bravo 2006 for a more measured assessment). This also suggests that it is William Junior who has made his mark rather than the more charismatic William Senior, who was famous during his lifetime but is now mostly remembered for the wrong reasons, as a butcher of whales whose suicide (imputed at the time to derangement) was unlikely to have been motivated out of 'any sense of guilt over the five hundred whales he had killed' (Hoare 2009: 287).

The most obvious memorial to the Scoresbys, however, lies neither in books nor in buildings nor in monuments, but in a particular place, first mapped by William Junior in the early 1820s and now bearing the family name, Scoresby Sound. Whether Scoresby Sound, an isolated fjord system in northern Greenland, is a fitting tribute to the Scoresby family is another matter. Colonized by Denmark in 1924, the area was populated in 1925 with Inuit families shipped in from over 1,000 miles away in a crude geopolitical effort to assert Danish sovereignty over north-east Greenland. Included among those moved were the grandparents of the late Danish Greenlandic visual artist, Pia Arke, who, after returning to her birthplace in the late 1990s, would go on to assemble an emotionally wrenching photo-history, published posthumously in 2010 as *Stories from Scoresbysund* [*Scoresby Sound*]. The book, like all of Arke's work, makes for uncomfortable reading, providing a stark reminder of the connections between Greenland's neocolonial present and its multiply inscribed colonial past (Huggan and Jensen 2016; Jonsson 2016). Folded into this history is the Scoresby name, which is remembered today even as the names of the Inuit who have lived for generations in the area are not. Arke's project is one of postcolonial counter-memory, ranged against those official versions of history that have effectively erased the presence of people whose lives are considered unimportant by those in positions of authority; considered unimportant, that is, unless they can be seconded to the needs, both real and imagined, of colonial power. These people, argues Arke, also have a right to be remembered. The current population of Ittoqqortoormiit (Scoresbysund) is difficult to ascertain, but recent census figures suggest a number of around 500. Arke's efforts aside, these 500 people's lives still remain largely unknown and undocumented. But at least they have not been taken from them, which is more than one can say for William Scoresby Senior's 500 whales.

Notes

1 The Edinburgh-based Wernerian Natural History Society (frequently abbreviated to Wernerian Society) was a nineteenth-century learned society named after the German geologist, Abraham Werner. It was founded and led by the Regius Professor of History at the University of Edinburgh, Robert Jameson, who was a

key patron and mentor for Scoresby Junior, and who – as the fulsome dedication to Jameson at the beginning of Scoresby's 1820 *Account of the Arctic Regions* clearly indicates – would exert a considerable influence on his work. For more on the relationship between Scoresby and Jameson, see Bravo 2006; also Scoresby-Jackson and Stamp and Stamp.

2 Standard critiques of Weber's work, both at the time and more recently, are that it is far too sweeping, effectively homogenizing both Protestantism (especially Calvinism) and capitalism while also paying scant attention to gender and promoting a culturalist understanding of capitalism that overplays its psychological implications at the expense of its material effects. For an early collection of critical essays, see Green 1959; for a more up-to-date assessment, see Lehmann and Roth 1993.

3 Mercantilism and imperialism are both much-debated terms, as is the relationship between them. The former has been defined in terms of a combination of 'controlled commercial expansion' and 'national identity and assertiveness', thereby creating (in the British context) a connection between 'a new pattern of economic development', generally associated with the Elizabethan era, and 'the rise of the British nation state' (Ormerod 2003: 3, 6; see also Wallerstein 1980 [vol. II]). This is far from an agreed-upon definition, however, and some imperial historians – notably John Darwin – have little time for mercantilist theories, although these have recently made something of a comeback in the work of, for example, Stern and Wennerlind (2013) and Phillip J. Smith (2015). Smith, in particular, sees mercantilism as a driving force behind British imperial expansion, although clearly this also depends on what is meant by the term 'imperialism', which is usually seen – albeit from a variety of sometimes conflicting perspectives – as *both* a theory *and* a practice of governance linked (in complex combinations that vary from time to time and place to place) to territorial expansion, cultural domination, the rise of industrial capitalism, and the consolidation of regional and/or national interests within a global system of international trade. For a useful overview of some of the issues, definitional and otherwise, at stake (focusing on the British context), see Darwin 2013 and Johnson 2002.

Whaling, as explained in my Preface, has been connected to both imperial and mercantilist interests. As Gordon Jackson argues (again in the British context), 'the prospect of a national whaling industry was consistent with the aspirations of the embryo mercantilists of Elizabethan England' (1978: 15), while for Philip Hoare it would later help build 'the foundations of empire' (2009: 277), not just by establishing 'new routes of oceanic colonization' (279), but by exerting influence in both the northern and, later, southern hemispheres (280) as well as creating and servicing a lucrative market in oil, among other whale-based commodities, that would see after-dark London by the mid eighteenth century glint and glitter as 'the best-lit city in the world' (277). That the Scoresbys were bound up in the cultural as well as the economic work of empire is indisputable, and while it would be too loose to describe either of them as an out-and-out imperialist, they both contributed to late eighteenth- and early nineteenth-century efforts to establish 'the northern ocean as a national space over which control [could be] increased by a range of military, commercial, regulatory, and rhetorical means' (Bravo 2006: 522). To this might be added their Enlightenment-inspired views on a range of subjects, including whaling itself, which they both saw as making 'an enlightened contribution to the nation' (Bravo 2006: 522), but which might be better considered in terms of a 'dialectic of Enlightenment' (Adorno and Horkheimer 1997 [1947]) that reveals the continuing will to dominate behind supposedly progressive ideologies of improvement and development: the dark heart of Enlightenment itself (see chapter epigraphs above).

4 In C.F Burgess's words, the 'fellowship of the craft', probably best described in Conrad's 1906 memoir *The Mirror of the Sea*, consists of 'the subtle, unspoken, and unseen tie which binds together all worthy seamen, each to the other, in dedication to the service of ships and the mastery of the sea' (quoted in Barrow Jr. 2015: 218; see also Burgess 1976). These bonds of solidarity are not as unequivocal as Burgess implies, being considerably complicated by issues of seniority, race, class, and indeed gender (e.g. different modes and models of masculinity) across the body of Conrad's work. Conrad still never quite manages to do what Melville excels at: upset and challenge the rigidly coded expectations of male behaviour that are the skewed accompaniment of a 'womanless world' (Baurecht 1986: 53). There is no space here to go into detail on the different forms of masculinity encoded in the life and work of the Scoresbys, both of which embrace many of the models critically examined by Raewyn Connell, who draws attention to (among several others) the figures of the 'imperial adventurer', which she also links to Weber, and 'the man of the frontier' (Connell 1995: 185, 187).

Similarly, there is no time to dwell on the broader connections between masculinity and the imperial imaginary, which remains a favourite topic of imperial history and postcolonial criticism, reflecting the fact that, as John Benyon puts it, possibly unaware of the double meaning, '[t]he [British] Empire generated a huge *demand for manpower* to serve its diverse needs, abroad at home, as leaders, administrators, clerks, soldiers, seafarers, merchants and so forth' (Benyon 2002: 26, emphasis mine). As discussed below, *northern* codes of masculinity also loom large in any assessment of the Scoresbys, whether in the context of the Arctic or that of northern England, although due attention needs to be paid to significant generational and temperamental differences between them, which are reflected in their recreational writing as well as their professional work.

5 There are of course as many different versions of the 'new man' as there are new men, each adapted to its particular historical context: hence the inadvisability of pushing comparisons too far between the Regency-style gentlemanliness of William Scoresby Junior and the more modern sensitivities, which are both a product of and a reaction against late Victorian/Edwardian social mores, of Edmund Gosse. Still, the fact remains that, separated though they are by the best part of a hundred years, they are both 'new men', if less by virtue of being in the vanguard of a new era than by inhabiting an uneasy space *between* eras. This, in Scoresby's case, can also be seen in terms of a transition between two different forms of *science* (involving a shift towards the professionalization of the scientific disciplines) and two different forms of *whaling* (involving a shift to more modern industrial methods of capturing and 'processing' whales).

6 Unsurprisingly for the time, the Scoresbys were both obsessed with finding a way to the North-West Passage, and a lively discussion of this possibility leads off the first volume of Scoresby Junior's *Account of the Arctic Regions*, which – in typical fashion – makes a rational case for the existence of the Passage, only to exercise a certain scepticism towards its own arguments: as Scoresby wryly remarks, northern voyages of discovery have provided 'a full illustration [of the maxim] that *what we wish to be true, we readily believe*' (Account, vol. 1: 3, emphasis Scoresby's). However, while both men were accomplished explorers, with Scoresby Senior sailing further north than any British navigator of his time, the twin goals of the North-West Passage and the North Pole remained elusive, with Scoresby Junior's mostly sensible ideas for reaching both destinations being repeatedly knocked back by the Admiralty (Fleming 1998; Stamp and Stamp 1976).

7 Davidson provides an entertaining long-historical account of the myths, for example those surrounding Ultima Thule, that have long swirled around the idea of north, which has been alternately imagined as a *profane* space, malign and destructive, and a *sacred* one: a region in which to purify and clarify the spirit and a place of beneficent transition in which high north, in particular, functions

as a kind of way station, a bridging territory between the world of the living and the world of the dead. See Davidson (2005: 47, 59, 64, 121); also Huggan (2016: 335).

8 It seems suitably ironic that right whales – at least in the Greenland Seas – were wrongly classified by the Scoresbys and their contemporaries. In the famous 'Cetology' section of *Moby-Dick*, Melville suggests that a working distinction might be observed between the 'Greenland whale' – which was also seen at the time as a right whale – and the 'American Right'. Melville was correct without knowing it: for the 'Greenland whales' that the Scoresbys were capturing and killing in their hundreds were in fact *bowheads*, a species related to but non-identical with the genus of the right whale. No small thanks to the Scoresbys, bowheads are almost gone now off the Greenland coast, although their numbers are healthy in other Arctic regions; however, what Melville calls 'American Rights' are in deep trouble, and rights more generally are among some of the most endangered whales on earth.

Works Cited

Abbs, Peter (1983) 'Introduction'. In Edmund Gosse (ed.) *Father and Son*. London: Penguin, 9–31.

Adorno, Theodor and Max Horkeimer (1997) [1947] *Dialectic of Enlightenment*. London: Verso.

Arke, Pia and Stefan Jonsson (2010) *Stories from Scoresbysund: Photographs, Colonisation and Mapping*. Copenhagen: Pia Arke Selskabet & Kuratorisk Aktion.

Barrow, Tony (2001) *The Whaling Trade of North-East England, 1750–1850*. Sunderland: University of Sunderland Press.

Baurecht, William (1986) 'To Reign Is Worth Ambition: The Masculine Mystique in *Moby-Dick*'. *Journal of American Culture* 9.4: 53–62.

Benyon, John (2002) *Masculinities and Culture*. Buckingham: Open University Press.

Bravo, Michael (2006) 'Geographies of Exploration and Improvement: William Scoresby and Arctic Whaling, 1782–1822'. *Journal of Historical Geography* 32: 512–538.

Burgess, C.F. (1976) *The Fellowship of the Craft: Conrad on Ships and Seamen and the Sea*. New York: Kennikat Press.

Burgess, C.F. (2015) 'The Fellowship of the Craft'. In Clayton R. Barrow, Jr. (ed.) *America Spreads Her Sails: U.S. Seapower in the 19th Century*. Annapolis: Naval Institute Press, 215–230.

Burnett, D. Graham (2012) *The Sounding of the Whale: Science and Cetaceans in the Twentieth Century*. Chicago: University of Chicago Press.

Connell, R.W. (1995) *Masculinities*. Cambridge: Polity Press.

Conrad, Joseph (2007) [1906] *The Mirror of the Sea*. Kindle Edition. CreateSpace Independent Publishing Platform.

Craciun, Adriana (2009) 'The Scramble for the Arctic'. *Interventions* 11.1: 103–114.

Crain, Caleb (2012) 'Melville's Secrets'. *Leviathan: A Journal of Melville Studies* 14.3: 6–24.

Darwin, John (2013) *Unfinished Empire: The Global Expansion of Britain*. London: Penguin.

Davidson, Peter (2005) *The Idea of North*. London: Reaktion.

Fleming, Fergus (1998) *Barrow's Boys*. London: Granta.

Garrison, Laurie (2008) 'Imperial Vision in the Arctic: Fleeting Looks and Pleasurable Distractions in Barker's Panorama and Shelley's *Frankenstein*'. *Science, Technology and the Senses* 52, http://id.erudit.org/revue/ravon/2008/v/n52/019804ar.html?lang=es, accessed 15 October 2016.

Gaskell, Elizabeth (1996) [1863] *Sylvia's Lovers*. London: Penguin.

Goldsmith, Oliver (1870) *Animated Nature*. London: Blackie and Son.

Gosse, Edmund (1983) [1907] *Father and Son: A Study of Two Temperaments*. London: Penguin.

Green, Robert W. (1959) *Protestantism and Capitalism: The Weber Thesis and Its Critics*. Boston: Heath.

Hill, Jen (2008) *White Horizon: The Arctic in the Nineteenth-Century British Imagination*. Albany: State University of New York Press.

Hoare, Philip (2009) *Leviathan or, the whale*. London: Fourth Estate.

Huggan, Graham (2016) 'Ultima Thule/The North'. In Carl Thompson (ed.) *The Routledge Companion to Travel Writing*. London: Routledge, 331–340.

Huggan, Graham and Lars Jensen (eds) (2016) *Postcolonial Perspectives on the European High North: Unscrambling the Arctic*. London: Palgrave Macmillan.

Jackson, Gordon (1978) *The British Whaling Trade*. London: Adam and Charles Black, Ltd.

Johnson, Rob (2002) *British Imperialism (Histories and Controversies)*. London: Palgrave Macmillan.

Jonsson, Stefan (2016) 'When Humanism Finds Its Ends: Lessons from Pia Arke and Katarina Pirak Sikku on the Difficulty of Narrating the Arctic'. In Lars Jensen and Graham Huggan (eds) *New Narratives of the Arctic*, special issue of *Studies in Travel Writing* 20.3: 226–236.

Lehmann, Hartmut and Guenther Roth (eds) (1993) *Weber's Protestant Ethic: Origins, Evidence, Context*. Cambridge: Cambridge University Press.

Lenz, Millicent and Carole Scott (eds) (2015) *His Dark Materials Illuminated: Critical Essays on Philip Pullman's Trilogy*. Detroit: Wayne State University Press.

Melville, Herman (2003) [1851] *Moby-Dick; or, the whale*. London: Penguin.

Morgan, Benjamin (2016) 'After the Arctic Sublime'. *New Literary History* 47: 1–26.

Ormerod, David (2003) *The Rise of Commercial Empires: England and the Netherlands in the Age of Mercantilism, 1650–1770*. Cambridge: Cambridge University Press.

Pratt, Mary Louise (1992) *Imperial Eyes: Travel Writing and Transculturation*. New York: Routledge.

Pullman, Philip (1995) *Northern Lights* (vol. 1 of *His Dark Materials)*. London: Scholastic Children's Books.

Pullman, Philip (2011) *His Dark Materials*. London: Everyman.

Rayment-Pickard, Hugh (2004) *The Devil's Account: Philip Pullman & Christianity*. London: Darton, Longman and Todd Ltd.

Richard, Jessica (2003) '"A paradise of my own creation": Frankenstein and the improbable romance of polar exploration'. *Nineteenth-Century Contexts* 25.4: 295–314.

Ryall, Anka, Johan Schimanski, and Henning Howlid Wærp (eds) (2010) *Arctic Discourses*, Newcastle: Cambridge Scholars Publishing.

Scoresby, William (1851) *Memorials of the Sea. My Father: Being Records of the Adventurous Life of the Late William Scoresby*. London: Longman, Brown, Green and Longmans.

Scoresby, William (2011) [1820] *An Account of the Arctic Regions*: vols. 1 and 2. Cambridge: Cambridge University Press.

Scoresby-Jackson, Robert Edmund (1860) *The Life of William Scoresby*. London: T. Nelson and Sons.

Smith, Phillip J. (2015) *The Rise and Fall of the British Empire: Mercantilism, Diplomacy and the Colonies*. Kindle Edition. Make Profits Easy LLC.

Stamp, Tom and Cordelia Stamp (1976) *William Scoresby: Arctic Scientist*. Whitby: Caedmon of Whitby Press.

Stern, Philip J. and Carl Wennerlind (eds) (2013) *Mercantilism Reimagined: Political Economy in Early Modern Britain*. Oxford: Oxford University Press.

Thwaite, Ann (2002) *Glimpses of the Wonderful: The Life of Philip Henry Gosse*. London: Faber & Faber.

Wallerstein, Immanuel (1980) *The Modern World-System*, vol. II, *Mercantilism and the Consolidation of the World-Economy, 1600–1750*. London: Academic Press.

Weber, Max (1974) [1930] *The Protestant Ethic and the Spirit of Capitalism*. London: Unwin University Books.

Whitehead, Hal and Luke Rendell (2015) *The Cultural Lives of Whales and Dolphins*. Chicago: University of Chicago Press.

Wilson, E.O. (1998) *Consilience: The Unity of Knowledge*. New York: Knopf.

Winter, Alison (1994) '"Compasses All Awry": The Iron Ship and the Ambiguities of Cultural Authority in Victorian Britain'. *Victorian Studies* 1: 69–98.

3

Killers: Orcas and Their Followers

[The killer whale] is the fiercest, most terrifying animal in all the world.

Joseph J. Cook

Dolphins never hurt anybody.

Richard O'Barry

The Shamu effect

In a 2005 story by the Canadian author Craig Davidson, a marine park trainer loses a leg after a spectacular routine with a performing orca, known in the trade as a 'rocket hop',[1] goes catastrophically awry. The accident is described in graphic detail:

> Niska's mouth opens. My left leg slips inside. Thigh raked down a row of teeth, shredding the wetsuit. Rocketing upwards, faster now. My crotch smashes the crook of her mouth and something goes *snap*. Jam a hand into Niska's mouth and pry with everything I've got, her jaws a jammed elevator I'm trying to open. Whale gagging on the foot lodged deep in her throat, huge muscles constricting and relaxing. Bubbles swirling and ears roaring, mind panicked and lungs starved for oxygen, a bright flame of terror dancing behind my eyes and yet there remains this great liquid silence, all things distant and muted in this veil of salt water. A disconnected image races through my head: that famous black-and-white snapshot of a Buddhist monk sitting serenely in lotus position as flames consume him. (Davidson 2005: 72)

The trainer in Davidson's story survives; several real-life counterparts have not been so fortunate. Take Keltie Byrne, for instance, who was dragged under and drowned by an orca in 1991 at the Pacific Northwest marine park Sealand: 'the first time anyone had been killed by a killer whale in captivity', writes the journalist David Kirby (2012: 12), but 'certainly not the last'. Byrne's killer was called Tilikum (Chinook for 'friend'): a huge, temperamentally unstable orca who would go on to kill two other people, including – some nineteen years later – the experienced and universally popular Orlando SeaWorld trainer, Dawn Brancheau. If Brancheau's death provided a poignant reminder of the risks involved in interacting with large predatory animals, the 'media frenzy' (Zimmermann 2010: 4) that followed would focus not on the dangers posed by orcas to humans but on the continuing exploitation of captive animals for entertainment purposes – an all-too-familiar morality tale of individual opportunism and corporate power (Davis 1997; Desmond 1999).[2] The negative publicity would eventually feed into the hit 2013 documentary *Blackfish*, a media spectacle of its own which has been credited with opening the eyes of many of its worldwide viewers to the cruelty of marine-mammal captivity – as well as the more specific animal-welfare abuses of the billion-dollar SeaWorld franchise – for the first time (Hargrove 2015; Kirby 2012). Litigation followed on both sides; activist groups sprang up; and several books and articles were published, some of them by former animal trainers turned full-time animal advocates who now felt suitably emboldened either to expose the rapacity and irresponsibility of their corporate employers (Hargrove 2015) or to reveal the error of their own ways (O'Barry 2012). Most of the action, however, has been on the internet, with numerous advocacy networks having being built up around social media,[3] and with individual celebrity animals – Tilikum included – having attracted large followings in an ironic, morally inverted rehearsal of the acquisitive tactics with which killer whales, particularly since the emergence of the corporate 'orca-display industry' (Neiwert 2015: 120) in the mid-twentieth century, have been pursued, rounded up and captured, and literally or metaphorically confined.

The size of this industry should not be underestimated. David Neiwert, in his 2015 study *Of Orcas and Men*, provides some startling statistics. In 2012, he says, 'orca facilities around the world drew over 120 million people, more than the combined attendance of Major League Baseball, National Football League, and National Basketball Association games' (121). Neiwert's analogies are apt, for since its inception in the 1960s, orca display – especially though by no means exclusively in the United States – has been a major spectator sport, a razzmatazz entertainment industry replete with loud music, bright lights, and crowd-pleasing acrobatics, bearing all the carefully choreographed characteristics of a 'nautical circus show' (227). Bearing all the risks, too, insofar as orca display, despite the innocent world it often seeks to represent, is an inherently dangerous business in which powerful animals – intensely vulnerable themselves – are coerced into performing alongside trainers whose commands are not always followed; whose human bodies are no match for those of their non-human counterparts; and whose staged attempts to interact with intrinsically unreliable, in some cases demonstrably dysfunctional animals relies on risk itself as a profitable commodity in the symbolic economy of human–animal exchange.[4]

One way of looking at the North American orca-display industry is through the lens of one of its earliest celebrity animal performers. As Jane Desmond has suggested (1999: 217), 'What Mickey Mouse is to Disneyland,

Shamu is to Sea World. Marketing symbol, ambassador, embodiment of dreams come true for children (and adults), Shamu, the most celebrated orca whale of all time, is the synecdoche of Sea World.' A direct physical and ideological correlation therefore exists between the staged personality of this engaging celebrity animal and the industry that would rapidly spring up around her (or, insofar as the Shamu brand has been able to benefit from a commercially exploitable blurring of gender boundaries, him: see Desmond 1999: 248). It thus seems fitting that Shamu, an origin myth of sorts, should be founded on a derived name, She-Namu, after her slightly earlier male counterpart, Namu, who though short-lived himself, has the best claim to being the first of the line.[5] Notwithstanding, the original Shamu, captured in 1965, was the first killer whale in SeaWorld's collection and soon became its tutelary spirit, its 'primordial goddess' (Hargrove 2015: 26). As John Hargrove – a former SeaWorld trainer, since turned star witness for the prosecution – sardonically observes, 'Every show was about Shamu; every whale at the center of the spectacle was called Shamu; any companion to the principal whale in the shows was somehow explained away as Baby Shamu or Great-Grandbaby Shamu. Shamu would never die. At least, not in name' (2015: 26). What I will call here the 'Shamu effect' depends, as Desmond convincingly contends, on promoting a utopian view of natural innocence that thinly masks American cultural dominance and corporate leadership (Desmond 1999: 217). It also relies on a strategic sidelining of violence, even as it is violence of different kinds – commercially derived, performance-bred, captivity-related – that makes the spectacle possible in the first place, and even as the risks associated with human–animal encounter, mediated by the heroic figure of the trainer, are made abundantly clear.

Above all, the Shamu effect is produced via what I have called elsewhere (2013: 4) the 'repeatable uniqueness' of the celebrity figure: here more specifically the iconic figure of the celebrity animal, which operates as both an ideological conduit for self-perpetuating 'family values' and an affective mirror for generally rose-tinted human perceptions of themselves (Blewitt 2013; Giles 2013). 'Repeatable uniqueness', of course, is by no means limited to animal figures, being an integral part of those wider media processes by which celebrities – which nearly always means human celebrities – are rendered simultaneously special and banal. This seemingly inexorable return to the human is typical of celebrity studies in turn, a field which has tended to focus almost exclusively on human subjects; and in which when animals have featured, they have usually been constructed as mute objects of human curiosity or doe-eyed recipients of self-serving philanthropic concern (Giles 2013). More recently, however, relations between humans and animals, and between the categories of 'human' and 'animal', have been made considerably more complex, in ways that theorists of celebrity – long fixated on the human – are only just beginning to understand (see Blewitt and Giles above; see also Boger 2015, Nance 2015).

Part of this understanding involves the acknowledgement of species difference. One of the ironies specific to animal celebrity is that it involves equal measures of anthropomorphism and dehumanization. Thus, while celebrity animals, like their human counterparts, are nothing without the (human) audiences that create them, unlike most of these counterparts they are the products of a coercive rather than voluntary performance in which they have little if any agential capacity to manipulate their stage roles or to fashion media-friendly images of themselves. David Giles thus gets it exactly wrong when he suggests that celebrity animals tend to 'wear the badge of authenticity held to be so important for credible image-management [since] there is never any question as to whether or not they are being themselves' (2013: 117). On the contrary, celebrity animals are *never* allowed to be themselves, and this is the most effective marker of their celebrity. (In the case of orcas, the ironies are compounded by the fact that they are not even allowed to belong to their own species; for though they are sometimes loosely affiliated with the *odontocetes* (toothed whales), killer whales are not usually classified as whales at all, but rather as the largest of the dolphins (*delphinidae*) [Ford, Ellis, and Balcomb 2000; see also Hoyt 1990 and Neiwert 2015].)[6] The Shamu effect provides a conspicuous example of this particular, media-driven form of ontological dissociation. 'Shamu' as stage name is transferred from one performing orca to another, not only implying that the number of Shamus is potentially unlimited but also suggesting that Shamu ultimately rests, not in the individual personality – albeit de-individualized persona – of any given performing animal, but in the involuntary nature of animal performance itself (Orozco 2013).

Not just animal performance, though, for the Shamu effect is produced through a *combination* of interactive performances: those of the orcas themselves, those of their trainers, and not least those of their family audiences, whose participation as well as appreciation is needed to complete the choreographed, almost balletic spectacle of controlled grace and power that provides the immediately recognizable signature of the live shows (Desmond 1999). The Shamu effect is thus the result of programmed interaction between multiple bodies organized around the iconic figure of the lead orca, whose 'physical body and its actions come to represent the complex of feelings, ideas, and fantasy that [constitute] the ideological subtext of the park' (1999: 222). It is crucial that the orcas are real, for although the Shamu effect is continually reproduced through the various simulacra (photographs, souvenirs, replicas, etc.) that are associated with the Shamu figure, only the direct physical presence of the orcas vouchsafes the authenticity of a performance that bears all the hallmarks – mass, power, danger – of 'the consumable sublime' (247).

More recently, specific forms of technological mediation, for example projection screens and digital imaging devices, have enhanced the general effect of the real by providing a more visually complex array that balances audience 'perception of the whales as the embodiment of physical difference and the abstraction and emblematization of that difference into graphic or schematic form' (247). To some extent, this turn to technology has been SeaWorld's less-than-subtle way of negotiating the fall-out generated by a sequence of accidents at its four marine parks, notably the death of Byrne, which resulted in the early 1990s in the cessation of all pool-based human–animal encounters, and with trainers since then being obliged to perform their routines either from poolside or from specially constructed shallow-water ledges that provide the possibility of a swift getaway should the orcas become aggressive and suddenly decide to attack (Hargrove 2015; Kirby 2012).

While the shows have become less spectacular as a result, they have also become more interactive in the sense that a wider range of participatory activities is now offered in which audience members are given the opportunity to fashion their own responses to what they are seeing, and to share those responses with others on social media, thereby creating their own individualized versions of the Shamu effect in which they themselves become celebrity performers – at least for a time. This is how celebrity has always worked: as a particular, attention-seeking kind of technologically mediated performance in which public performers (celebrities) depend upon, and are to some extent created by, their followers, who then – with varying degrees of success – fulfil their fantasy of becoming public performers themselves (Gamson 2011; Rojek 2001; Turner 2004). But as Sharon Marcus has recently argued, the opportunities provided by new media, while by no means eliminating the status hierarchies that attach to older forms of celebrity, have made it significantly easier for 'fans to address celebrities, celebrities to address fans, and fans to address one another' – and for fans to *become* celebrities, even if few ever do and celebrity still remains 'an exclusive status reserved for a very few people, a status that many people imagine they would like to possess but [most of them acknowledge] they won't obtain' (Marcus 2015: 49). Notwithstanding, recent evidence suggests that there is an increasing democratization of celebrity that has similarly closed the gap between human and animal celebrities, while the rapid development of new technologies has also effected a transition from the human/animal celebrity as fetish object – which theorists such as Kelli Fuery see as being characteristic of 'old' media – to the fetishization of celebrity status, which may be conferred upon celebrities, celebrity-followers, and follower-celebrities alike (Fuery 2008: 139; see also Driessens 2015).

The rest of this chapter focuses on the various followings – both virtual and real – that have accompanied recent developments in the North American orca-display industry, organizing its thoughts around the celebrity-animal figures of Tilikum, Keiko, and Morgan, the last of these currently held at one of Europe's few remaining marine parks to keep orcas in captivity, Loro Parque, on the popular North Atlantic holiday island of Tenerife.[7] All three emerge as embattled figures whose celebrity has come at a price, and whose life stories provide sad confirmations of the continuing human exploitation of performing animals, much of which is tied in with the depredations of the modern corporate world (Wilson 2015). They also – especially Tilikum – afford reminders of the modern celebrity industry's fetishization of death, revealing what John Blewitt calls the morbidly 'shamanistic quality [that] can be detected in the morphology of celebrity presentation and reception', whether the celebrity scenario in question is human, animal, or – as is nearly always the case with animal scenarios – a densely tangled combination of both (Blewitt 2013: 330; see also Rojek 2001: 53–56).

Performing orcas, whose scientific name (*Orcinus orca*) links them to the classical underworld,[8] are perhaps uniquely positioned to permit insight into the 'dark side' (Kirby 2012) of the late twentieth- and early twenty-first-century marine-mammal entertainment industry, which brings together at least three different kinds of celebrity – celebrity *commerce*, celebrity *activism*, and celebrity *conservation* – each of which draws substantially on the others while critically reflecting on them at the same time (Brockington 2009; Huggan 2013). As I will go on to show, this particular celebrity scenario – even as it trades on the innocence of its animal and, more occasionally, human victims – is threaded through with the half-wished-for threat of violence, offering further opportunity for critical reflection on the linked forms of fanaticism (obsession) and moral recoil (obscenity) that are at the dark heart of celebrity itself.

After *Blackfish*

Gabriela Cowperthwaite's powerful documentary film *Blackfish* – named after the popular Pacific Northwest Native American/Canadian term for the orca, a revered ancestral figure in many of their myths and legends – premiered at the Sundance Festival in 2013, and was a more or less instant hit, gathering increasing attention as one of the most talked-about documentary movies of the year (Neiwert 2015: 232). Although the film only brought modest box-office returns, it would find a much larger audience on cable television, and a still larger one via such popular web-based outlets as iTunes and Netflix (Brammer 2015: 74). Most of all, *Blackfish* – ostensibly an animal advocacy vehicle loosely organized around the Tilikum attacks and the increasingly desperate attempts of SeaWorld to protect its own commercial interests – became and has remained a major catalyst for hard-fought internet campaigns both for and against marine-mammal captivity, so much so that the Shamu effect has spawned a *Blackfish* equivalent that has reached tens of millions of homes and placed the entire marine-park industry on red alert (Brammer 2015; Hargrove 2015; Neiwert 2015).

This latter effect has registered on several different scales, from specific SeaWorld stock depreciation – especially in the immediate aftermath of the film – to overall declining marine-park attendance, and it continues to be fought out in several different arenas that join political strategy to legal conflict, in the process exposing serious fault lines and rivalries within what only superficially seems to be the same moral cause.[9] But by far its most conspicuous front has been that of social media, with a continuing battle being played out online, both through organized opponents' official websites and, just as important, through spirited (sometimes spiteful) individual exchanges between different impromptu participants in what might be seen at one level as an unofficial national conversation and, at another, as a major global public debate. As Rebekah Brammer observes, participation in such debates 'is not so much about what you read as what you share, retweet, tag, like, comment on or blog about – and users don't have to like or follow an organisation in order to do this' (Brammer 2015: 74). And as she further suggests, the battle, joined this way, is an example not just of 'how the source medium – the film itself – has been able to reach a far bigger audience through social media', but also of how social media themselves fuel the expectation of both individual and collective cross-platform participation, meaning that 'we can no longer assess the success of a film through the more traditional models of box-office gross and DVD sales alone' (78).

What *Blackfish* foregrounds as well is the prevalence in our times of web-based forms of social and environmental activism that offer the utopian possibility of transforming 'emotion into action' (Castells 2012: 13), even if they do not *necessarily* turn into action, and even though they may effectively *substitute* emotion for action, as several recent popular critiques of so-called 'keyboard activism' attest (Brammer 2015; Chiaramonte 2012). One view of cyberactivism – those different forms of social and political protest which find their primary means of expression on the web – is that it has effectively changed what counts as activism, paving the way for new kinds of 'decentred' social movements that are more open and democratic than the older ones they seek, whether explicitly or implicitly, to replace (Castells 2012). A less rosy view is that cyberactivism is a by-product of 'digital capitalism' (Barassi 2015) and as such shares many of its most salient characteristics: that is, it is almost constitutively unreliable, and therefore easy to co-opt by less-than-progressive commercial interests; and it tends by definition to be temporary, if by no means inconsequential, in its social and political effects (McCaughey and Ayers 2003).

As the example of *Blackfish* shows, both of these views have some truth to them. Thus, a prominent (not to mention predictable) feature of the online forums is the large number of factual mistakes in them or, more to the point, the large number of times that 'facts' are drawn from the film either to support individual opinion or to lend credence to only one, often extreme, side of the debate (Kleiman 2014). Another, related feature is the use of social media (including almost inevitably the 'big three', Facebook, Twitter, and YouTube) to protect as much as to assail corporate interests, in this case those of SeaWorld, which maintains a carefully managed website (www.seaworld.com) that draws repeated attention to the company's latest animal-welfare initiatives and its ongoing conservation work. As Brammer points out, communications on both sides, even as they are widely differentiated and dispersed, are often framed in binary terms (sometimes via shortened identifiers such as 'pro-cap' versus 'anti-cap'), which allow moral victories to be proclaimed at different moments while regular insults – accompanied by equally regular charges of defamation – are traded across the bows (Brammer 2015: 75). Such high-octane exchanges are routine on the web, as are the media-friendly forms of individualization and personification used to identify them. Thus it should come as no surprise to learn that alongside numerous organized activist initiatives (The Orca Project, Orca Aware, Voices of the Orcas, etc.), there are several Tilikum-specific Facebook pages (Tilikum the Orca, Retire Tilikum the Orca, Free Tilikum), which attract a wide variety of loosely monitored postings, from short vituperative statements to longer, more considered blogs.

Nor should it be a surprise that celebrities have joined the fray, thereby grabbing the chance to act as trendsetting moral arbiters in debates which, at their most emotional, revolve around unashamedly anthropomorphized issues of animal personhood and slavery, with individual celebrity animals – Tilikum prominent among them – being used as identifying markers for the larger conservationist cause (Brammer 2015; Kirby 2012). Most of *Blackfish*'s celebrity followers have unequivocally supported the film, with popular TV actors from Aaron Paul (*Breaking Bad*) to Ricky Gervais (*The Office*) in turn attracting large fan followings, while musicians have also been particularly active in their condemnation of SeaWorld, with various North American recording artists – Barenaked Ladies, Cheap Trick, and Willie Nelson among them – demonstratively pulling out of previous bookings to play at one or another of the company's marine parks (Brammer 2015: 76).

As Dan Brockington asserts, 'The flourishing of celebrity conservation is part of an ever-closer intertwining of conservation and corporate capitalism' (Brockington 2009: 2). Celebrities have long since voiced public support for a variety of conservationist initiatives, using them both to boost their own moral credentials – a media-induced manoeuvre archly described by Max Boykoff and John Goodman (2009: 396) as 'conspicuous redemption' – and to feed off conservation's own media-conscious, often corporation-supported, manipulations of 'affective power' (Marshall 1997: xii). In the case of *Blackfish*, three interrelated kinds of affective power are made evident: the power of the documentary genre, the power of the charismatic animal, and the power of celebrity conservation itself. Let me take each of these briefly in turn here. The success of documentary, Bill Nichols suggests, depends on its capacity to reinforce viewers' 'preexisting emotional attachments to representations' as a means of winning them over to the argument it constructs (Nichols 1991: 135). *Blackfish* organizes its own particular emotional appeal around the figure of the orca as a large charismatic animal whose intelligence and physical force are rendered proportionate with its suffering, and whose immediate recognizability, together with its symbolic resonance, make it a particularly attractive candidate for financial support (Brakes 2009). The disproportionate amount of conservation funding given over to 'charismatic megafauna' has been well documented, as has the media fixation on so-called 'flagship species', with both of these potentially distracting attention away from more important ecological players, for example keystone species, and more inclusive forms of biodiversity conservation and ecosystem management (Leader-Williams and Dublin 2000: 54). 'Celebrity conservation' is the composite term that best describes the different ways in which conservation initiatives draw strength, both from specific celebrity figures and from the general 'celebrity system' (Marshall 1997: xii) within which these figures operate – a system which frames celebrity as a commodity, subject to the fluctuating rates and regimes of value that dictate symbolic as well as material exchange (Brockington 2009; Huggan 2013).

In *Blackfish*, animal advocacy provides a paradigmatic example of the workings of the celebrity system, bringing together different celebrities (both human and animal) and different *kinds* of celebrity (both achieved and attributed) within the overarching context of the global conservationist cause.[10]

For if Tilikum is the undisputed star of the show, doubling – as celebrities often do – as hero and villain, he also shares space with several other celebrity figures, Dawn Brancheau prominent among them as another media-hounded victim, chased first (and killed) by Tilikum, then by a legion of public sympathizers – from investigative journalists to self-identifying followers and supporters – seeking to make sense of her death. While the success of *Blackfish* has been to engage these sympathies at large, it is the internet rather than the film itself that has mobilized them in even greater numbers by providing 'multiple outlets for the personalised creation of celebrity cultures, celebrity products, celebrity selves' (Huggan 2013: 12). What is characteristic about much of this cross-platform activity is its high degree of individualism, from the personal nature of the responses, some of which claim an intimate relationship to the victims they cannot possibly have,[11] to its focus on charismatic individuals, with both of these in keeping with the celebrity system's restless fashioning of 'manufactured personalities', who come to embody 'the spirit of individualism in contemporary consumer culture [as well as fulfilling] the material and ideological requirements of the system that creates them' (Marshall 1997: 247). To some extent this is a function of contemporary animal advocacy's strategic emphasis on the welfare of individual animals (especially suffering animals) whose personhood is directly linked to sentience, with strong (and often strongly anthropomorphic) forms of transpersonal identification, then further strengthened by the medium, being the most visible result (Beers 2006; Kemmerer 2015). To some degree it is a function as well of the individualization of politics that is sometimes seen as being one of the most notable by-products of cyberactivism – a process that depends not only on strong emotional connection to particular causes but also on the self-publicizing technologies that allow such emotions to be made visible and expressed (Castells 2009; Meikle 2002). Most of all, however, it is a function of *visibility itself* as the implied goal of such activities, which are part of what Joshua Gamson among others sees as a 'heightened consciousness of everyday life as a public performance: an increased expectation that we are being watched [and a corresponding] willingness to offer [our private selves up] to watchers known and unknown' (Gamson 2011: 1068; see also Marcus 2015, Turner 2004).

In his hard-hitting analysis of the media-driven controversies that *Blackfish* helped create, Joe Kleiman (2014: 210) points out two major influences on the way the information surrounding them has been disseminated: the rise of social media and the spread of personalized digital imaging technology, both of which make instant news without necessarily giving much attention to the accuracy of what they report. Kleiman, who admits to being an interested party, is too keen on clearing SeaWorld's name, mainly by directing his ire at what he sees as the 'tabloid' tactics of PETA (People for the Ethical Treatment of Animals) and other leading animal-advocacy organizations, but also by listing a number of deliberately 'misleading edits' in the film, which he considers as manipulating the evidence in favour of the anti-captivity argument it presents (10, 14, 33–35). In so doing, he mistakes the whole purpose of documentary film, which, as Nichols notes, is expressly designed to win consent for a particular viewpoint (Nichols 2001: 4), while remaining silent about his own position, which tacitly supports the view of display animals as 'assets' in an industry the financial interests of which apparently need more protection than the animals themselves (Kleiman 2014: 8).

Still, Kleiman scores several palpable hits (factual inaccuracies, opinion masquerading as evidence, etc.) against the film. He also shows how at least some activist organizations have made political capital out of it, and perhaps above all how the self-publicizing environment in which new media often operate tends to turn collective protest into individual performance and to promote the very consumerist ideologies it claims to contest.[12] As I have been suggesting thus far, this environment requires both a redefinition of celebrity in terms of the performative capacity to attract a *following* and a critical re-assessment of the different situated meanings of 'following' itself. This makes the orca-display industry an interesting case study, not just because it involves an intersection of public performances that are framed by celebrity images but also because it plays between real and virtual followings, framing these in turn in terms of accumulation (following as collective noun) and pursuit (following as active verb). Such followings are rarely if ever innocent. In the case of orcas, the stakes are high and following involves at least four different if closely connected kinds of violence: the violence of capture, the violence of captivity, the violence of representation, and the more occasional violence of orcas themselves. The more particular case of Tilikum brings all four kinds of violence to the fore while also gesturing towards what Marcus (2015: 49) calls the 'resonant paradoxes' of celebrity culture, with the giant orca bearing upon his battered body[13] the alternate patterns of heroism and demonization that have historically been inscribed onto the Janus-faced figure of the killer whale (Hoyt 1990; Neiwert 2015). A rather different case, albeit one that has attracted an equally large following, is that of Keiko, the eminently marketable narrative of whose capture, rehabilitation, and release would provide the basis for the saccharine 1993 Warner Bros. movie *Free Willy*, but which contains – as I will now go on to elaborate – not one but several, significantly less attractive, backstories of its own.

Killing Keiko, freeing Morgan

Keiko – the 6000-pound male orca with a female name – shares the stage today with Tilikum as the world's most famous killer whale, and although the reasons for his celebrity are different, at least some of the many stories that continue to swirl around him are the same. Keiko, like Tilikum, was captured in Icelandic waters, where, also like Tilikum, he was snatched from his family and sold to a North American marine park, proceeding to endure rough treatment as a calf made to share a tank with several older and more aggressive killer whales (Kirby 2012: 197–198). The uninterrupted object of media attention from that time (the early 1980s) right up until his death (in the early 2000s), Keiko is a celebrity in all the classic senses of the word. Celebrities, I have argued elsewhere, are 'discursively produced through media and other communications networks; they are symptomatic of the blurring of private and public spaces in everyday social life; they are brand names and marketing tools as well as cultural icons and model identities; and they are both targets of and vehicles for a wide variety of cultural and ideological debates' (Huggan 2013: 1). Keiko would fulfil all of these functions and more in the context of a short life in which he was both syndicated film star (*Free Willy*) and international tourist magnet – media darling, too, with the various news features that surrounded his rehabilitation, his eventual release, and the persistent controversies that accompanied them, helping to create mass followings that have only expanded further in the 'cyberactivist' internet age (McCaughey and Ayers 2003).

In fact Keiko seems always to have been followed in one way or another, whether by hunters and collectors associated with the orca-display industry, or by advocates and activists committed to the animal liberation movement; whether by North American and Mexican marine-park visitors during his time in captivity, or by Icelandic and Norwegian tourists during his time in the wild.[14] In the process, he has been tracked over thousands of miles, relocated and re-released, and subjected to a wide range of commercial and scientific interests. Little wonder that Kirby, who has in turn tracked the large number of often competing activities that involve him, describes him ironically as 'the perfect whale' (2013: 199) – the multiply desired object of intertwined public-relations machinations, fan activities, and press interests (Giles 2013: 124), both celebrated for his freedom yet, like other celebrity animals, eternally confined.

Many of the activities surrounding Keiko had come together by the mid-1990s in the so-called 'Keiko Project', a loose amalgam of animal-advocacy organizations arranged around a broadly liberationist platform which, as things turned out, some individual members would choose not to support. In the eye of the storm was Mark Simmons, the ex-SeaWorld trainer hired in 1998 to help rehabilitate Keiko, who had recently been flown to his new Icelandic home, a sea pen in the Westman Islands, where for the first time in two decades he could experience the relative freedom of ocean life (Kirby 2012: 262–263). Simmons, along with another former SeaWorld employee, Robin Friday, was quick to denounce the same rehabilitation programme he had been appointed to direct, leading to ugly recriminations about SeaWorld – which maintained a keen interest in Keiko, though from a pro-captivity standpoint – having effectively taken over the Project with a view to making it fail (Kirby 2012: 264). The Keiko Project, Simmons and Friday maintained (though at least some of the evidence suggested the opposite), was not cut out for success. Nor for that matter were a number of other proposed cetacean-release programmes, which were making the same mistake of imagining that captive animals, once liberated, would adapt quickly and un-problematically to the wild (Neiwert 2015: 213–221).

Simmons would later leave the project, but return with a vengeance with a 2014 book, *Killing Keiko: The True Story of Free Willy's Return to the Wild*. The book's pro-captivity agenda is clear, as is its provocative ambition to show that, far from helping Keiko, the various animal-advocacy organizations involved in his rehabilitation were at least partly responsible for his death. There is scant evidence to support this view – Keiko would eventually die of pneumonia, a relatively common cetacean disease, with little to suggest that this had been brought on by his experiences in the wild. But the book opportunistically rides the storm that surrounded his death, which would draw a number of angry protests about the huge and unnecessary expenditure of the Keiko Project; its celebrity-obsessed diversion of funds away from other, more deserving collective animal-rescue projects; and, in some of the more extreme cases, its wrong-headed insistence on the very possibility that captive animals could re-adapt to the wild (for a summary of responses, see Kirby 2012: 279–281).

Simmons's book, which seemed part-designed to settle old scores, may have been ill advised while – as numerous reviewers were quick to point out – it was also scientifically inaccurate. But it was effective nonetheless in drawing attention to the factional tensions within the contemporary animal-advocacy movement, which Diane Beers has wryly described in terms of a smorgasbord of sometimes only vaguely related platforms ranging in size and species from Save the Dolphins to United Poultry Concerns (Beers 2006: 200). And it was effective, too, in showing the centrality of the media, both 'old' and 'new', to the movement, as well as the importance of manipulating public relations in the service of an environmental cause. The Keiko Project, in fact, might well be construed as being more about the dark arts of PR than about the mistreatment of animals in captivity, though at a time when followings can be accumulated at the click of a button, PR – still understood primarily in terms of the corporate maintenance of a favourable public image – involves a multidirectional process that combines many different publics, many different public identities, and many different, often highly personalized, conceptions of what a 'public image' is.[15]

More recent variants of the Keiko Project, such as the 'Free Tilikum' and 'Free Morgan' campaigns, rely to an even greater extent on social media as a versatile tool for the privatization of public relations as well as a multifunctional instrument for the mobilization of ethical concerns. The latter campaign is spearheaded by the Free Morgan Foundation (www.freemorgan.org), a non-profit charity set up in 2011 to provide information on, and solicit funding for, an ongoing court case based on the proposed repatriation of Morgan, a young female orca currently held at a Tenerife-sited marine park, Loro Parque, to the wild. Some years later, the case, which has already gone through numerous appeals, remains unsettled, although a 2015 ruling failed to establish grounds for repatriation and Morgan, whose health has visibly suffered in captivity, remains caught between the courts (in the Netherlands, where she was originally rescued) and her far-from-pristine artificial surroundings (in the Canary Islands, which belong to Spain). The campaign, as one might expect, features a variety of individualizing tactics, in some of which Morgan talks 'directly' to us, and along with the usual fund- and attention-raising mechanisms (donations, petitions, etc.), the website has an active Twitterfeed (8000+ followers) and Facebook account (30,000+ likes). As might also be expected, its expert board consists mostly of scientists, a few of whom were also involved in the Keiko Project, while listed supporters include HRH Prince Albert of Monaco, the *Born Free* actress Virginia McKenna, and three Dutch supermodels, 'Morgan's Angels'.

This is no Keiko Project, then, but rather a fairly standard small-scale operation aimed at building an international cross-platform following, and using social media as its primary means of doing so. Unashamed about its own commercialism, the campaign represents – like many of its kind – a working compromise between cross-species advocacy, which registers a general 'commonality of ethical concern for animals' (Beers 2006: 4–5), and species-specific activism, which uses the case of one particular celebrity animal to agitate for major industry transformation and, beyond it, grassroots social and political change. These activities, taken together, show the close interworkings between celebrity and conservation, an alliance that consolidates 'the capitalist system within which both [of these] are materially embedded by tying the exchange value of threatened species to that associated with celebrity itself' (Huggan 2013: 250). But they also hint at the well-documented tensions between individually oriented animal rights and systems-based environmentalist perspectives that can be found in many media-driven global conservation movements, with celebrity intervention in these movements usually tending to accentuate the former at the expense of the latter, as wide-ranging studies from Dan Brockington's (2009) to Lisa Kemmerer's (2015) show.

Kemmerer's work in particular illustrates that the focus of animal-based celebrity activism is on suffering, with the possibility of death a constant spectral presence hovering over both liberationist and protectionist debates. There has always been an element of voyeurism in the Western cetacean imaginary, which further embellishes the already extreme violence inscribed within the history of human–cetacean encounters (Hoare 2009; see also Chapter 1). But in the case of the orca, this violence has less to do with its material exchange value as an outsize bundle of commodities (as is the case with, say, the sperm whale) than with its symbolic status, itself thoroughly commodified, as a 'killer' whale. As Desmond notes, there is something obscene about the way in which the violence of the orca-display industry is assiduously moved off stage, so that when it does make its occasional public appearance, there is a kind of forbidden pleasure in the spectacle, 'as obscene as the sight of pristine azure pools polluted with blood' (Desmond 1999: 249). The Davidson passage with which I began this chapter brings out the semi-pornographic nature of a trainer–orca encounter that goes horribly wrong, producing multiple forms of forbidden exposure: the exposure of secret complicity, the exposure of animal violence, the exposure of human flesh. This obscene scenario, laced with pleasurable violence, is also at the heart of the case study with which I will close, another Tilikum-induced fatality, this time involving the curious figure of Daniel Dukes.

Conclusion; or, the killer whale as obscene object of desire

It is still far from clear why, on 5 July 1999, the 27-year-old drifter Daniel Dukes broke under cover of night into SeaWorld Orlando's Shamu Stadium, whereupon he proceeded to strip down to his trunks and jump into G Pool for an ill-advised swim with its temporary occupant, Tilikum the killer whale. What is clear, though, is what happened to him. When Dukes was found dead the next morning, slumped over a now-placid Tilikum, there were multiple wounds all over his body, his crotch and left leg were flayed, and his left testicle had been almost surgically removed from the scrotum, with divers having to retrieve it from the bottom of the pool (Kirby 2012: 259). Unsurprisingly, the media were quickly onto the story, with both 'pro-cap' and 'anti-cap' versions of it circulating widely, most of them understandably blaming Dukes but a good few pointing out the psychotic tendencies of Tilikum, and some of the more lurid construing the encounter as a bizarre meeting of human and animal delinquents, one perhaps secretly seeking and the other savagely delivering death (259–260; see also Hargrove 2015: 98).

Still stranger versions circulated on the internet, with one story speculating that Dukes, a fanatical animal-lover, might have looking to act out his fantasies by having sex with a killer whale. This sensationalist combination of physical violence and sexual innuendo was almost certainly one of the motivating factors behind Davidson's aforementioned story 'Rocket Ride', even though the orca in the story bears more resemblance – at least in name – to Kiska, Canada's sole remaining captive killer whale. It also played on one of the thematic preoccupations of Davidson's work, namely the obscene, which is typically defined in terms of the physical (usually sexual) excesses that are brought to the surface because, rather than in spite, of the fact that they are being officially curtailed. As Kerstin Mey observes, obscenity is not inherent in an object or event, but is rather 'an argument about [its] qualities, public exposure and traffic' (Mey 2007: 2). This makes the case of Dukes an object lesson in the obscene, although an equally convincing case can also be made for the official Shamu shows, which link the routine atrocities of circus display to those of corporate capitalism, and in which the different forms of structural violence involved in orca captivity (capture, breeding, training) become apparent even as they are consistently denied or de-emphasized, literally or metaphorically shifted off stage. As Mey (2007: 1–2) points out, obscenity, which is derived from the Latin *ob scaena* (off stage), usually involves the displacement of an event or object that is not intended for public display, the revelation of something (or someone) that has deliberately been hidden. Obscenity is thus a form of moral disgust, but simultaneously a registering of illicit desire for that which is officially found to be disgusting. In the context of the orca-display industry, this disgust translates into a publicly expressed outrage against the accumulated abuses of what Hargrove (2015: 8, 108) melodramatically calls 'a rapacious corporate scheme that [has] exploited both the orcas and their human trainers', turning the former into a cross between 'gladiator-slaves' and 'performing prisoners'. Meanwhile, no less floridly, Kirby (2012: 169) describes the SeaWorld corporation as 'a slow-motion death machine for killer whales'.

Obscenity is perhaps a word too easily used, but the violence is real enough, whether or not it is owned up to, and it has not lost its capacity to attract the same mass followings it shocks. Recent news from SeaWorld is mixed. At

SeaWorld San Diego, the Shamu shows are to be phased out, in keeping with the corporation's possibly disingenuous pledge to move away from circus tricks towards educational imperatives (Watson and Schneider 2015). At the same time, SeaWorld continues to battle the courts for the right to hold cetaceans in captivity, while at its San Antonio franchise, a captive dolphin has died – the third of the marine park's animals (the others were an orca and a beluga whale) to die in a space of three months (Kaplan 2016). In all of this, it is difficult not to see the figure of the orca – particularly the celebrity orca – as an obscene object of desire for its followers, whether they are actively involved in or resolutely opposed to the entertainment industry that surrounds it, and whether they doggedly support or energetically campaign against the continuing captivity of killer whales. Perhaps this is one, eminently sad, version of the fate of the celebrity animal, which, while ostensibly serving to bridge the divide between human beings and their non-human others (Blewitt 2013: 336), ends up either widening it or translating it into potentially pathological scenarios in which non-human killers are followed, and human followers are killed.

Coda: After Tilikum

On 6 January 2017, Tilikum died at 36 years of age, 33 of those spent in captivity (Konstantinides 2016). As I have observed elsewhere (Huggan 2013), nothing becomes a celebrity in his (or her) life like the leaving it. Celebrity animals are no exception, and Tilikum's death attracted a storm of mostly negative publicity, with numerous opportunities being taken to reiterate the cruelties of cetacean captivity and the inalienable rights of all cetaceans to 'life, liberty, and well-being' (Marino 2014: 33). Some commentators pointed out that, at 36, Tilikum was still relatively young, confirming a long-standing pattern of premature death in captivity, while some went further to offer punishing reminders of other orcas languishing in inadequate conditions at US and European marine parks (Anon. 2016a and b). One such orca is Morgan, who while still alive is hardly well, and who has recently been showing increased signs of stress, including a much-publicized 2016 incident where she 'beached' herself for ten minutes on the concrete at the side of her tank. Loro Parque administrators – not for the first time facing a PR disaster – were quick to explain this away as part of her 'normal' post-performance routine, but few industry analysts, and fewer still animal activists, were convinced by their account.

Not long after the 'beaching' incident, Ingrid Visser and Rosina Lisker published a lengthy report with further damaging evidence of cetacean mistreatment at Loro Parque (Visser and Lisker 2016). The report focuses initially on the animals' poor dentition, which is mainly caused by the raking of teeth on gates or the hard sides of tanks, a key element of cetacean stress physiology that has been amply demonstrated elsewhere (see, for example, Marino 2014: 31). The report goes on to cite the inappropriate use of the medical tank, the lack of proper shade, and an apparent violation of SeaWorld's ban on captive breeding (SeaWorld technically owns all of Loro Parque's orcas, but this has not prevented Loro Parque from challenging SeaWorld protocols and from continuing the earlier, since discredited style of circus-entertainment shows). Itemized behaviour is then provided for each of the six orcas: in Morgan's case, this includes lunging for food (which is potentially dangerous for trainers and orcas alike), stress-related concrete-chewing (which helps account for the appalling state of her teeth), stereotypical pattern swimming and head bobbing (generally acknowledged as further signs of stress), and some evidence of injury caused by other orcas (which suggests that Morgan since her arrival at Loro Parque in 2011 has never been fully integrated, and that sexual and other forms of aggression towards her are probably on the up).

The report takes care not to single Morgan out, also mentioning Loro Parque's other five captive orcas, but unsurprisingly it is Morgan – the celebrity among them – who continues to capture most media attention. This is no bad thing of course if, as is often the case with celebrity figures, one of the positive functions of celebrity status is to act as a conduit for potentially industry-changing public debates (Huggan 2013; Turner 2004). However, despite the good works of the Free Morgan Foundation, which also financed the report, there is little sign yet that Loro Parque is ready to succumb to the public pressure that has overtaken its larger sister organization, SeaWorld. And if, as Lori Marino concludes in her passionate account of the abuses that continue to be done to cetaceans in captivity, there are indeed signs that the marine-mammal industry is on the wane, the dismal prospect remains that it will be the likes of Morgan and Tilikum, rather than the entertainment industry which exploits them, that end up 'follow[ing P.T.] Barnum into the void' (Marino 2014: 33).

Notes

1 For many years, the 'rocket hop' was the *pièce de résistance* of SeaWorld shows, a culminating pool-based routine in which the trainer, balanced on the pectoral fins of a leaping orca, would be propelled thirty feet or more into the air, with both human and animal performers executing a perfectly coordinated back dive. Such routines, which included the 'stand-on' and the 'hydro', have since been discontinued, with even SeaWorld supporters such as Joe Kleiman admitting that choreographed shows in which orcas feature as little more than circus animals need to be consigned to the past (Kleiman 2014: 10).

2 Jane Desmond's brilliant analysis of the SeaWorld orca shows, while it predates Brancheau's death and the significant industry changes it enforced, has not been bettered. As Desmond suggests, the shows, while designed to reflect the 'horizontal' family values of affection, respect, and trust, were also carefully edited to avoid a 'visible show of force, aggression, or violence of any kind' (Desmond 1999: 218). However, the 'vertical' lines of command were clear, as was the utopian political model on which the shows were founded, which positioned America as 'benevolent patriarch' in a consensually domesticated world (250).

3 The Orca Project is one of the more prominent networks, advertising itself on its state-of-the-art website as a 'small but effective' non-profit organization whose self-appointed task is to 'change the public's attitude and government supervision of marine mammals in captivity through research, investigation and education'. While the project's role is ostensibly a monitoring one, its Facebook page (registering around 3600 likes at the latest time of writing [April 2017]) mostly features an array of overtly anti-captivity comments, suggesting that many of its users see themselves as requiring no further education, having already made up their minds about where they stand.

4 SeaWorld has repeatedly insisted on the safety of its orca shows, although various highly publicized court cases – notably OSHA vs. SeaWorld (for detailed accounts, see Hargrove 2015 and Kirby 2012) – have brought to light glaring inadequacies in relation to the living conditions of its marine mammals, especially if not exclusively orcas, and the training regimes of its staff. That the orca-display industry has been an extended exercise in risk management is crystal-clear – hence SeaWorld's huge and continuing investment in public relations – but it is equally clear that its potential dangers have been part of its attraction and have contributed to its success.

5 Namu, named after the remote fishing town in British Columbia where he was captured in 1965, quickly became a star turn as part of the Seattle Marine Aquarium franchise, gaining a celebrity status that only escalated when he featured in the uninspiring, but nonetheless popular, 1966 movie *Namu, the Killer Whale*. Namu himself, however, would not outlast his fame, succumbing first to a bacterial infection then, scarcely a year after his capture, getting caught in the net designed to contain him, where he drowned (Kirby 2012: 151–152; Neiwert 2015: 109–111).

6 Technically, orcas (sometimes used in the plural without the 's') are not a single species but a 'species complex'. This complex incorporates a range of diverse populations, usually divided into 'resident' and 'transient' varieties, both predominantly coastal, along with a smaller number of offshore whales (Neiwert 2015: 73–82). That orcas are killers is not in dispute, although they are rarely if ever aggressive in the wild towards humans – not that the same thing can be said for the other way round (Moe 2014).

7 Most European countries no longer hold orcas in captivity, the exceptions being France and Spain, with Loro Parque – a particular target of anti-captivity campaigns, many of them organized around the female orca Morgan – holding all of Spain's six orcas, although the ownership issues around them are anything but resolved. Although Morgan's parentage is disputed (she was originally rescued in Dutch coastal waters), she continued until recently to be claimed by SeaWorld as part of a captive breeding programme – now discontinued – that proved to be every bit as controversial as its shows (Kirby 2012; see also Marino 2014). At the latest time of writing (April 2017), SeaWorld holds 22 orcas at its US marine parks, while twice as many have previously died in captivity. Recent evidence suggests that Russia, China, and Japan may be looking to expand their own orca-display industries even as most of their global national counterparts contract (Neiwert 2015).

8 *Orcinus orca* is derived from Orcus, one of the Roman gods of the underworld (Neiwert 2015: 99). As David Neiwert points out, there is a stark contrast between the ways in which Western (e.g. Greco-Roman) myths portray orcas and the reverential treatment accorded to them by different indigenous cultures, from the Yupik in Siberia to the Haida in the Pacific Northwest (98–99).

9 The highest-profile cases have been those involving the US Occupational Safety and Health Administration (OSHA), which since the mid-2000s has repeatedly levied fines – since upheld by the courts – against SeaWorld for its violation of professional safety requirements; and those involving the pro-animal-rights

NGO, People for the Ethical Treatment of Animals (PETA), which filed a 2011 lawsuit against Sea World contending that the corporation was in violation of the 13th Amendment as five captive orcas – including Tilikum – were effectively slaves. PETA's case was lost as these orcas were not found to be legal 'persons' and were thus not liable to protection under the terms of the US Constitution, but the case continues to be discussed widely and has a lengthy media-sustained afterlife of its own. Not all animal-advocacy groups have sided with PETA, though in rubbishing PETA's views, Joe Kleiman (2014) caricatures an entire anti-captivity movement which, in demonstrating a common concern for the welfare of animals, has historically been fractured around definitions and interpretations of animal rights. For more on the ethical issues surrounding cetaceans in captivity, see Marino 2014; on a more general level that includes considerations of cetacean 'personhood', see also Chapter 2 and Chapter 4.

10 Chris Rojek's distinction between 'achieved' and 'attributed' forms of celebrity has been influential, although as Rojek himself admits, the lines between separate categories of celebrity can easily become blurred. 'Achieved' celebrity, Rojek suggests, is legitimate fame based on demonstrable talent or skill in a particular area, whereas 'attributed' celebrity is based on the more general ability to generate public interest, for example by using different media outlets to create some kind of following, whether or not that following is deserved (Rojek 2001: 19). Animals further complicate these categories insofar as their achievements are themselves attributed (by humans); as John Blewitt emphasizes, animals are usually given celebrity status because they have the capacity to act as 'attractors' that allow human beings to reflect upon themselves (Blewitt 2013: 328).

11 So-called 'para-social relations' (Turner 2004: 92) with celebrities have a long history, although their effects may be intensified by new media, which offer a multiplicity of indirect or surrogate forms of contact with people who are, to all intents and purposes, unknown. Para-social relations are generally understood to be one-sided exchanges where interest in a particular person is not reciprocated, leading to obsessive–compulsive behaviour that is usually harmless, but may occasionally have pathological results (93).

12 As Graham Meikle argues, cyberactivism in some cases is little more than a form of 'libertarian hype' which, far from registering dissent against corporate authority, provides a further reminder of the continuing 'corporate colonisation of cyberspace' (Meikle 2002: 39).

13 Despite his formidable reputation during his time in captivity, Tilikum was roughly treated as a calf – an all-too-familiar tale that also applies to Keiko and Morgan,

and which almost always happens when captive calves, separated from their parents, are made to share tank space with older, non-family-related killer whales.

14 As Kirby notes, not only was Keiko 'tracked by satellite all the way across the Atlantic' after his release, but he was then pursued by tourists at his eventual chosen home in Skålevik Fjord (Kirby 2012: 270). Here 'he splashed and posed for photos with the good people of coastal Norway', some of whom had 'boated or even [swum] out into the inlet to play with the Hollywood celebrity' (Kirby 2012: 270–271). Neiwert probably says it best when he claims, citing support from cetacean scientists, that 'Keiko's story vividly demonstrates how *not* to return an orca to the wild' (Neiwert 2015: 220). It is less clear what *would* have been a good way to do it given the inevitable attention given to celebrity animals, who are by definition destined to remain captive products of the followings their circumstances create.

15 As Edwards Bernays, sometimes dubbed 'the father of public relations', observes, public relations are not primarily about publicity, but about 'efforts to integrate [the] attitudes and actions of an institution with its publics and of publics with [that] institution' (Bernays 2013 [1952]: 3). Bernays recognizes in the process that such 'adjustments' (his term) often involve the alignment of public with private interests; however, what we see increasingly today – not least through the workings of new media – is a blurring of public and private realms. Public relations, in this sense, are as likely to be in individual as in collective hands, with individuals being able to some extent to create their own publics – a phenomenon that can be seen in certain, hyper-individualized forms of media activism in which individuals are arguably less interested in manipulating public opinion about institutions than in manipulating public opinion about themselves.

Works Cited

Anon. (2016a) 'Over 30 Years and Three Deaths: Tilikum's Tragic Story'. www.seaworldofhurt.com/features/30-years-three-deaths-tilikums-tragic-story, accessed 20 April 2017.

Anon. (2016b) 'SeaWorld Orca Tilikum That Killed Trainer Dies'. www.bbc.co.uk/news/world-us-canada-38531967, accessed 20 April 2017.

Barassi, Veronica (2015) *Activism on the Web: Everyday Struggles against Digital Capitalism*. London: Routledge.

Beers, Diane (2006) *For the Prevention of Cruelty: The History and Legacy of Animal Rights Activism in the United States*. Athens: Ohio University Press.

Bernays, Edward (2013) [1952] *Public Relations*. Norman: University of Oklahoma Press.
Blewitt, John (2013) 'What's New Pussycat? A Genealogy of Animal Celebrity'. *Celebrity Studies* 4.3: 325–338.
Boger, Peter C. (2015) 'The Nature of Celebrity and the Celebrity of Nature: Digital Adaptation and Wildlife Survival in the Age of the Anthropocene'. Madison: University of Wisconsin. Unpublished PhD dissertation: publication no. 3707807.
Boykoff, Max and John Goodman (2009) 'Conspicuous Redemption? Reflections on the Promises and Perils of the "Celebritization" of Climate Change'. *Geoforum* 40: 395–406.
Brakes, Philippa (2009) 'Are Whales and Dolphins the Great Apes of the Oceans?' us.whales.org/wdc-in-action/are-whales-and-dolphins-great-apes-of-oceans, accessed 1 September 2015.
Brammer, Rebekah (2015) 'Activism and Antagonism: The *Blackfish* Effect'. *Screen Education* 76: 73–79.
Brockington, Dan (2009) *Celebrity and the Environment: Fame, Wealth and Power in Conservation*. London: Zed Books.
Castells, Manuel (2009) *The Rise of the Networked Society: Economy, Society, and Culture*, vol. 1. Oxford: Wiley-Blackwell.
Castells, Manuel (2012) *Networks of Outrage and Hope*. Cambridge: Polity Press.
Chiaramonte, Perry (2012) 'Slacktivists: Changing the World with "Likes", Clicks and Tweets?' *Fox News*, April 13. http://www.foxnews.com/tech/2012/04/13/slacktivists-changing-world-with-likes-clicks-and-tweets, accessed 1 September 2015.
Cowperthwaite, Gabriela, dir. (2013) *Blackfish*. New York: Magnolia Pictures.
Davidson, Craig (2005) 'Rocket Ride'. In *Rust and Bone: Stories*. Toronto: Viking Canada, 69–100.
Davis, Susan (1997) *Spectacular Nature: Corporate Culture and the Sea World Experience*. Berkeley: University of California Press.
Desmond, Jane C. (1999) *Staging Tourism: Bodies on Display from Waikiki to Sea World*. Chicago: University of Chicago Press.
Driessens, Oliver (2015) 'The Democratization of Celebrity: Mediatization, Promotion and the Body'. In P. David Marshall and Sean Redmond (eds) *A Companion to Celebrity*. Oxford: Wiley-Blackwell, 371–384.
Ford, John K. B., Graeme M. Ellis, and Kenneth C. Balcomb (2000) *Killer Whales: The Natural History and Genealogy of Orcinus Orca in British Columbia and Washington State*. 2nd ed. Seattle: University of Washington Press.
Fuery, Kelli (2008) *New Media: Culture and Image*. Houndmills: Palgrave Macmillan.
Gamson, Joshua (2011) 'The Unwatched Life Is Not Worth Living: The Elevation of the Ordinary in Celebrity Culture'. *PMLA* 126.4: 1061–1069.
Giles, David C. (2013) 'Animal Celebrities'. *Celebrity Studies* 4.2: 115–128.

Hargrove, John (2015) *Beneath the Surface: Killer Whales, SeaWorld, and the Truth Beyond Blackfish*. London: Palgrave Macmillan.

Hoare, Philip (2009) *Leviathan or, the whale*. London: Fourth Estate.

Hoyt, Eric (1990) *Orca: The Whale Called Killer*. Willowdale: Camden House.

Huggan, Graham (2013) *Nature's Saviours: Celebrity Conservationists in the Television Age*. London: Earthscan/Routledge.

Kaplan, Sarah (2016) '3 Animals Have Died in 3 Months at SeaWorld San Antonio'. *Chicago Tribune*, February 12. http://www.chicagotribune.com/news/nationworld/ct-seaworld-san-antonio-deaths-20160210-story.html, accessed 12 February 2016.

Kemmerer, Lisa (ed.) (2015) *Animals and the Environment: Advocacy, Activism, and the Quest for Common Ground*. London: Routledge.

Kirby, David (2012) *Death at SeaWorld: Shamu and the Dark Side of Killer Whales in Captivity*. New York: St Martin's Griffin.

Kleiman, Joe (2014) 'Dissecting Blackfish'. micechat.com/wp-content/uploads/2014/01/dissectingblackfishfinal.pdf, accessed 1 September 2015.

Konstantinides, Anneta (2016) 'The Cost of Captivity: Orca Blackfish Star Tilikum – Who Killed Seaworld Trainer – Dies, Aged 36, after 33 Years in a Tiny Tank after Cousin in the Wild Reached Ripe Old Age of 105'. www.dailymail.co.uk/news/article-4095148/SeaWorld-Tilikum-orca-killed-trainer-died.html, accessed 20 April 2017.

Leader-Williams, Nigel and Holly T. Dublin (2000) 'Charismatic Megafauna as "Flagship Species"'. In Abigail Entwhistle and Nigel Dunstone (eds) *Priorities for the Conservation of Mammal Diversity: Has the Panda Had Its Day?* Cambridge: Cambridge University Press, 53–81.

Marcus, Sharon (2015) 'Celebrity 2.0: The Case of Marina Abramović'. *Public Culture* 27.1: 21–52.

Marino, Lori (2014) 'Cetacean Captivity'. In Lori Gruen (ed.) *The Ethics of Captivity*. New York: Oxford University Press, 22–37.

Marshall, David (1997) *Celebrity and Power: Fame and Contemporary Culture*. Minneapolis: University of Minnesota Press.

McCaughey, Martha and Michael D. Ayers (eds) (2003) *Cyberactivism: Online Activism in Theory and Practice*. London: Routledge.

Meikle, Graham (2002) *Future Active: Media Activism and the Internet*. Sydney: Pluto Press.

Mey, Kerstin (2007) *Art and Obscenity*. London: I.B. Tauris.

Moe, Peter Wayne (2014) 'Sounding the Depths of the Whale'. *Interdisciplinary Studies in Literature and Environment* 21.4: 858–872.

Nance, Susan (2015) *Animal Modernity: Jumbo the Elephant and the Human Dilemma*. Houndmills: Palgrave Macmillan.

Neiwert, David (2015) *Of Orcas and Men: What Killer Whales Can Teach Us*. New York: Overlook Press.

Nichols, Bill (1991) *Representing Reality: Issues and Concepts in Documentary*. Bloomington: Indiana University Press.

Nichols, Bill (2001) *Introduction to Documentary*. Bloomington: Indiana University Press.

O'Barry, Richard (2012) *Behind the Dolphin Smile: One Man's Campaign to Protect the World's Dolphins*. San Rafael, CA: Earth Aware Editions.

Orozco, Lourdes (2013) *Theatre and Animals*. London: Palgrave Macmillan.

Rojek, Chris (2001) *Celebrity*. London: Reaktion.

Simmons, Mark A. (2014) *Killing Keiko: The True Story of Free Willy's Return to the Wild*. New York: Callinectes Press.

Turner, Graeme (2004) *Understanding Celebrity*. London: Sage.

Visser, Ingrid N. and Rosina B. Lisker (2016) 'Ongoing Concerns Regarding the SeaWorld Orca Held at *Loro Parque*, Tenerife, Spain'. www.freemorgan.org/visselisker-2016-ongoing-welfare.concerns, accessed 20 April 2017.

Watson, Julie and Mike Schneider (2015) 'SeaWorld to End Orca Shows in San Diego by 2017'. *Associated Press*, November 9. www2.kqed.org/news/2015/11/09/seaworld-to-end-orca-shows-in-san-diego-by-2017, accessed 12 February 2016.

Wilson, David A. H. (2015) *The Welfare of Performing Animals: A Historical Perspective*. New York: Springer.

Zimmermann, Tim (2010) 'The Killer in the Pool'. *Outside Online*, July 30. www.outsideonline.com/1924946/killer-pool, accessed 1 September 2015.

4

Kind of Blue; or, the Infinite Melancholy of the Whale

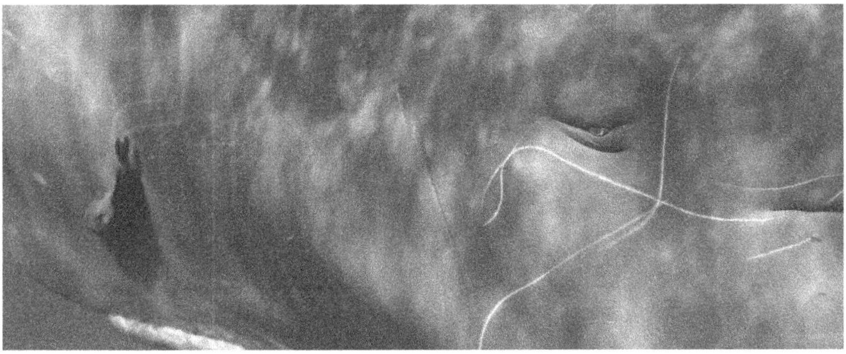

What happens to beasts can happen to man. All things are connected. If the great beasts are gone, men would merely die of a great loneliness of spirit.

Indian Chief Seattle, quoted in Frank Stewart, *The Presence of Whales*

Whales tend to lodge in the breasts of people – sometimes they lodge crosswise, where they stick for life.

Cetologist Roger Payne, quoted in Frank Stewart, *The Presence of Whales*

Whale/Fall

Whales' eyes have no tear ducts: we do their crying for them. Marine biologists are sometimes moved to tears at the sight of a whale, creative writers and historians by the thought of one. Why is that whales have the capacity to move us so, pitching us between alternating states of hopelessness and elation? The answer I will give in this chapter is that whales are compound figures for *melancholy*, itself a plural state that has historically been construed as veering between the two extremes of despair and exaltation, conjuring up utopian possibilities of transcendence but also spiralling down into agonized reconfirmations of the inevitability of death and loss. Whales – as already examined in Chapter 1 – have historically served as reversible signifiers of salvation and perdition; long associated with human stories of planetary annihilation, they are also standard bearers for the promise of human survival and the possibility of a 'better', more ecologically harmonious life.

It is this very possibility that connects whales with melancholy, making the destructive history of human encounters with whales that much more poignant. In a similar spirit, this last chapter will circle back to the first, offering some scattered reflections on whales as projection screens for human desires and interests, but also as spectral figures whose literal as well as figurative elusiveness ends up troubling the process of representation itself. For, as previously argued, it is the *spectrality* of the whale – simultaneously substantial and insubstantial – that helps turn it, not just into a quintessentially unsettling figure, but also into an all-purpose symbol for entangled histories of disappearance and loss.

Furthering this argument, my main contention in this chapter will be that the attribution of melancholy to the figure of the whale is less an effect of observable animal behaviour (ethology) than the projection of a human mood (sadness) together with an overriding guilt based on historical knowledge of the human exploitation of whales (see also Preface). A finer-grained version of this contention will need to account for melancholy as at once a composite of associations relating to loss and fear and a broad-based cultural symptomatology of conflicting emotions whose outwards signs are made manifest in the visible grandeur of the whale. Whales, in this last sense, can be seen to operate under the sign of *excess*: they are not just bodies but gigantic bodies, not just signs but a plethora of signs; and insofar as the long history of their material destruction and industrial disassembly – which Nicole Shukin (2009) ingeniously associates with the double meaning of the verb to 'render'[1] – has been accompanied by the proliferation of images that surround them, whales also become figures for *allegoresis*: not just a profusion of allegories, but the making of allegory itself.

To put this differently, whales are associated with the knowledge of loss, but also with wider social and cultural perceptions of loss tied in with the depredations of imperialism and industrial capitalism. Whales, in other words, are spectral or uncanny figures in the broad sense that their fleeting, potentially fateful appearances, replicated today in the more ostensibly benign practices of whale watching, operate as a kind of filter for the return of the repressed. Or, to phrase it differently again, it is the limited visibility of whales to the human eye that invites speculation on the scale – the enormity – of the losses that surround them; these losses are then translated into the representational codes of allegory, dolefully defined by Richard Stamelman as 'the language of fragmentation, decay, and erosion which death speaks or writes' (Stamelman, quoted in Pick 2011: 76).

For Stamelman, allegory represents the symbolic yoking of 'an absent and unrecoverable meaning' to 'an excessive and overdetermined language' (quoted in Pick 2011: 76). This – I want to argue here – is at one and the same time the language of *melancholy*, which as Max Pensky among others suggests has less to do with the experience (real or imagined) of loss than with a paralysing perception of the impossibility of recovery (Pensky 1993: 17–18; see also Radden 2000, Santner 2006). Melancholy, for Pensky, is a dialectical 'way of seeing' (Berger 2008) in which the inevitable price paid for penetrative insight is pain or anguish; it is also a debilitating habit of thought in which contemplative reflection effectively preys upon itself (19). Pensky links this self-destructive 'way of seeing' to an antiquarian perception of history as a repeating series of catastrophes that reaches back into the ancient past (18).

Meanwhile, Eric Santner – taking his cue, as Pensky does, from Walter Benjamin – links it further to the folding of *human* history into *natural* history, interpreting the latter in terms of 'the ceaseless repetition of […] cycles of emergence and decay of human orders of meaning [that] are always connected to violence', and to the inherent vulnerability of what he calls, again closely following Benjamin, 'creaturely life' (Santner 2006: 17). The figure of the melancholy whale is closely associated with the idea of creaturely life, which homes in not just on the intense vulnerability of the victimized human/animal body but on what Patricia Lopez and Kathryn Gillespie chillingly call the 'economic logics of killable life' (Lopez and Gillespie 2015; see Armstrong 2008, Butler 2010, Pick 2011).The figure of the melancholy whale is also linked to its status as an *affective animal*: an involuntarily public creature, both unsolicited object of media attention and opportunistic catalyst for media-generated activist and/or advocacy work. Affective animals are so named because they are able to elicit public sympathy, for example through what the Swedish media theorist Birgitta Höijer calls those forms of 'emotional anchoring' that activate the 'recognition of "our" complicity in "their" plight' (Höijer 2010: 719, 723–724; see also Nyman and Schuurman 2016). In Höijer's own words, 'emotional anchoring' refers to 'those communicative processes by which a new phenomenon is attached to well-known positive or negative emotions: for example fear or hope' (79). These anchoring processes are complicated, however, by the uncertain relationship between sympathy and understanding in affective animals. For instance, the capacity of animals to affect us isn't necessarily dependent on how much we understand them; indeed, our sympathy may be the greater the *less* we know about them. In this context, melancholy might be seen as registering the split between sympathy and knowledge, which reinforces a sense of the separation between us and other living creatures. In the case of whales, this separation is merely increased by our historical awareness of the suffering caused to these creatures, which involves a degree of personal identification that counteracts what Lauren Berlant calls the 'political soothing' of pain (Berlant, quoted in Nyman and Schuurman 2016: 67).

This personal identification may be experienced subjectively or projected onto something/someone else; similarly, it may be historically based or future-oriented, as in the apprehension of vulnerability as the imaginative foreshadowing of future pain. Whatever the case, it is the co-existence of sympathy and separation that is key to the social construction of the melancholy animal. Here, Laura Brown is quite right to see the melancholy animal as the by-product of socially, culturally, and historically differentiated instincts of sympathy and compassion, together with a deep sadness at the apparently degenerated state of the world (2010: 29). She is wrong, though, to claim that animals are 'either absolutely alien or intimately familiar' (16) to humans; for in the case of the whale – and the same is largely the case for other affective animals – it is clear that attributed melancholy is the joint effect of *both*. Thus, whales and other cetaceans may be perceived at the same time as being like us – 'the humans of the ocean', in Kalland's sardonic phrase (2012: 19) – and profoundly unlike us; as inhabiting our world and as living in a radically different, and indeed ontologically incommensurable, realm. Similarly, imaginative attempts to represent whales often widen the very gap they wish to narrow, reconfirming the distance that separates them from humans, so that what we are left with is a series of (1) tantalizing glimpses and (2) equally partial traces of 'actual animals within human texts' (Brown 2010: 24).

To sum up my argument so far, through the figure of the melancholy whale, melancholy can be understood as an *affect* rather than an *emotion* produced by instinctive feelings of sympathy accompanied by the painful recognition of separation.[2] This then shifts into full-blown melancholia – melancholy's pathological state – when these feelings become more permanent, and when separation anxiety transforms into a kind of precognitive state of mourning for originary loss. As Jennifer Radden argues, melancholia is the afflictive condition under which melancholy ways of seeing and thinking are normalized even as they take on pathological effects and characteristics: when our perception of our separation from others – be they human or animal – turns into a paralysing apprehension of our separation from ourselves (Radden 2000: 17–18).

As I argued earlier (see Chapter 1), this overriding sense of loss is bound up with the threat of species extinction. However, while it is true that some varieties of cetaceans have disappeared for good, the vast majority of them are still with us; and if, as Dan Bortolotti puts it, the figure of the melancholy whale registers, not just mourning for previous death, but guilty recognition in face of the apparently dying, then that presumed death sentence has turned out to be premature (2009: 91). Perhaps it might be better, in the current context, to speak of whales in terms of *pseudo-extinction* rather than *extinction* narratives. In the classic extinction narrative, there is mourning for actual loss, although that mourning may take several different forms, each suffused with cultural meaning (Heise 2016; Van Dooren 2014). In the pseudo-extinction narrative, by contrast, species may be *imagined* as being lost even though the science tells us otherwise.[3] The much-debated Freudian distinction between mourning and melancholia comes to mind here. Whereas mourning is predicated upon the identification of a lost object, melancholia is unable to locate it; is uncertain in fact whether the object is actually lost or not. The temporal dimension of melancholia is equally unclear, for whereas mourning is by definition projected onto the past, melancholia may just as easily be projected onto the future, anticipating a future state that not only assumes the worst but also narcissistically accepts responsibility for it (Radden 2000).

One possible analogy to this is the Judaeo–Christian doctrine of the Fall, the original state of sin from which mankind may eventually be delivered, but which may paradoxically involve some form of ultimate and/or irreversible judgement, for example the total destruction of the world. As is well known, there is a long-standing mythological connection between the figure of Leviathan and the presentiment of the End, in which the utopian promise of a new order already contains within it the seeds of its own destruction (see also Chapter 1). In the melancholic imaginary, the next rise is but the beginning of the next fall, confirming the hopelessness of messianic promise. This may be seen in political as well as religious terms; as Pensky puts it, 'Melancholy infects any political cause that it seeks to support, for the melancholic's support is always tinged with the atmosphere of meaninglessness. [...] A melancholy politics thus glides into a secret, half-willing collaboration with the forces it seeks to oppose' (1993: 11).

A remarkable literary demonstration of melancholy politics is László Krasznahorkai's award-winning novel *The Melancholy of Resistance* (1998), set in a provincial town in an unstable eastern European country (which both is and is not the author's native Hungary) subject to the absurd violence that routinely accompanies autocratic power. Loosely based on Robert Burton's seventeenth-century treatise *The Anatomy of Melancholy*, the novel plays out a series of mock-apocalyptic rehearsals, as the town's beleaguered citizens anxiously anticipate a triumphant deliverance that never comes. Instead, the novel cynically confirms its protagonist Valuska's vision of history as an endless repetition of errors and failures: bad things happen, worse things happen, and then they happen all over again (1998: 80–81, 104). At the centre of it all is a travelling circus and its prime exhibit, 'the biggest whale in the world' (26), a tawdry throwback to nineteenth-century spectacles presided over by the equally unprepossessing figure of the Director, himself 'an enormous behemoth of a man' (85). The giant stuffed whale comes across in the novel as an essentially incomprehensible figure: 'It was just too big and too long: Valuska simply couldn't see it all at once, and failed even to get a proper look at its dead eyes' (88). The creature is significantly less interesting than the promise it holds – typically, either salvation or perdition (see Chapter 1) – but above all, it comes to demonstrate 'an infection of the imagination whose susceptibility to its own terrors might eventually lead to an actual catastrophe'

(96): a combination of the preposterous and the monstrous, it embodies the affliction of melancholy itself. A suitably dingy signifier of apocalypse, the whale is a melancholic figure in a novel stuffed full of melancholic figures whose resistance to authority turns out to be as senseless as the wanton exercise of authority itself.

Krasznahorkai's melancholy whale is a good working example of Pensky's view of melancholy as an expression of political frustration as well as personal unhappiness – an expression he sees as being exemplified in Walter Benjamin's work. For Pensky, Benjamin is the epitome of the melancholy thinker, gifted with profound insight but tortuously aware that 'misery is the force that drives the contemplative mind onward' (Pensky 1993: 17). He is also a melancholy writer insofar as melancholy writing operates 'under the sign of its own impossibility to achieve what it needs' (1993: 5; see also Chambers 1987). Last but not least, he is an eagle-eyed analyst of the capacity for self-delusion that accompanies even the most worked through of revolutionary politics, and of the temptation to lapse into a pessimism so extreme that it effectively impedes the possibility of political change (1993: 9). In this context, *The Melancholy of Resistance* offers a set of wry reflections on the irresistible tendency of melancholy towards self-parody, its self-destructive privileging of thought over action, and its signal inability to resist – and secret complicity with – the very totalitarian forces of authority it confronts.

Krasznahorkai's most obvious targets here are the socialist regimes of twentieth-century eastern Europe, which he views as replicating the fascism they oppose, but his novel opens up to other, still more general interpretations, implying the whale-like hollowness of all political credos as well as the equally whale-like invitation to be swallowed by visions of an alternative order that proves to be just as corrupt and self-serving as the one it promises to supplant. All grand ideas, Krasznahorkai tacitly suggests, run the risk of turning into something like their opposites, or of degenerating into mere shells of themselves that demonstrate the one true melancholic principle: the certainty of decay. Hence the extraordinary ending to the novel, which consists of an extended, meticulously technical account of the *post mortem* process of bodily putrefaction (in this case, the body belongs to another of the novel's hapless victims, Mrs Jószef Plauf). This account out-Flauberts Flaubert in its rapt attention to the brute physiology of decomposition, offering one last, supremely ironic example of the futility of resistance as the rotting corpse, now ground down into 'infinitesimal pieces', returns to the chaos from whence it came (Krasznahorkai 1998: 314).

An equally arresting example of this process can be found in Rebecca Giggs's short 2015 piece in *Granta*, 'Whale Fall', which documents – in excruciating detail – the final agonies of a beached whale. The whale, a Western Australian humpback, takes a full three days to die, during which time it becomes a public spectacle. The initial festivities, however, give way to a stunned contemplation of the prolonged death process, in which the author and a steadily diminishing circle of other interested bystanders both witness the whale's obvious suffering and transfer it, in classic melancholic fashion, onto themselves. Apocalyptic rumours attach themselves like limpets to the body of the dying whale:

> Spectators on the beach that day had their suspicions. A comet had flared icily over Rottnest Island the previous week. The much-adored Antarctic Blue whale skeleton in the Western Australian Museum in Perth was set to be lifted out of the roof on a crane and dismantled. (At long last, would it be returned to the sea?) Someone's sister talked of significant and recent naval operations. And wasn't the weather undeniably weird all the time now? An El Niño year. Bitter mention was made by many of 'The Japanese'; of the trauma and exhaustion of whales chased by harpoons. Almost certainly, it was said by one man, the Nynungar elders predicted the humpback – that's why they weren't there. What was happening was something sour, something dark. A bad business for the land. (2015: 5)

Meanwhile, the whale itself begins to disintegrate as some organs fail, others move alarmingly from inside to outside the body, and the starving organism, 'breaking down blubber into energy in the absence of food, [...] reverts to ketosis', effectively poisoning itself from within (3, 7). There is still time, though, for Giggs to imagine one more whale death, way out to sea, where the whale body, sinking and decaying as it sinks, 'reaches a point where the buoyancy of its meat and organs is only tethered down by the force of its falling bones' (8). The description that follows is as emotionally draining as it is scientifically accurate:

> Methane is released in the miniscule bubbles. It scatters skin and sodden flesh below it, upon which grows a carpet of white worms (grass on its grave). Then, sometimes, the entire whale skeleton will suddenly burst through the

> cloud of its carcass. For a time, the skeleton might stay hitched jerkily at the spine to its parachute of muscle; a macabre marionette dangling in the slight currents. Then it drops, falls quickly to the sea floor, into the plush cemetery of the worms. Gusts of billowing silt roll away. The mantle of the whale's pulpier parts settles over it. Marine snow (anonymous matter, ground to a salt in the lighter layers of the sea) beats down ceaselessly. Rat-tails, devouring snails and more polychaetes appear. The bones are stripped and then fluff up with silver-white bacteria, so that it appears as if the skeleton is draped in metres of downy towelling. Years may pass, decades even, before there is nothing left except a dent that holds the dark darker. (8–9)

Giggs's is an accomplished piece of melancholy writing, in which the multiple associations of the Fall – both literal and figurative, both physical and metaphysical – are used to contemplate the mysteries of creation as well as the natural histories of destruction (natural history *as* destruction) that link terrestrial human beings to their alternately familiar and alien marine counterparts, whales. In Giggs's work, we see melancholy writing's putting into effect of a 'burden of remorse' as well as its working through of a 'perturbed consciousness'; we also see the melancholy state as a kind of 'eternalization of pensiveness' in which the melancholy thinker/writer is forever 'propelled forward, never ceasing [his/her] mulling over of things' (Chambers 1987: 31, 169, 173). Above all, though, we see these thoughts and their written expressions projected onto the figure of the dying (or deceased) *whale*, the literal dissolution of which makes way for a plurality of figurative associations that come together in the simultaneous recognition of sympathy and separation, the self-perpetuating contemplation of loss and the dizzying imaginative possibilities of what Judith Butler (2010) calls 'grievable life'.[4]

These associations, for Giggs as for Krasznahorkai, are mainly in the visual realm, although – as might be expected with the spectral figure of the whale – they continually play between the visible and the invisible, revealing what Pensky calls the 'structural ambiguity [...] between meaning and meaninglessness, life and death, exaltation and despair [that lies] at the heart of [melancholic] allegoresis as nothing other than an insight into the very structure of imagination itself' (1993: 28). However, the melancholic imaginary is stimulated as much by sounds as by sights, by auditory as by visual impulses; and in the next section of this chapter, I will go on to explore what probably remains the most popular association of whales with melancholy: the haunting music of humpback song.

Whale/song

Few sounds are more iconic than those of humpback song, which, to many of those who first heard it in the 1960s, confirmed that '[n]ot only were the whales on their way to extinction, they were singing their own dirge' (Ellis 1991: 436; also quoted in Bortolotti 2009: 88–89). As we now know, the extinction of cetacean species has largely been forestalled, although several varieties remain listed as threatened, but the haunting sounds of humpback song still continue to have a strong hold on the public consciousness, gracing many a nature documentary and forming the principal subject of several popular-science books, notably the American musician and philosopher David Rothenberg's acclaimed 2008 monograph, *Thousand Mile Song: Whale Music in a Sea of Sound*. For Rothenberg among others, the smash-hit record *Songs of the Humpback Whale*, first released in 1970 and later re-released as a 'sound page' in a 1979 issue of *National Geographic*, would do more than anything or anyone else to bring the plight of whales to public notice: 'We would never have been inspired to save the whale', he says unequivocally, 'without [having been] touched by its song' (Rothenberg 2008: 3). And if whale song was key, in the late twentieth century, to kick-starting a 'Save the Whales' campaign that would later become metonymic for the global environmental movement, it is increasingly acknowledged that acoustic research, in the early twenty-first, 'may turn out to be the single most useful tool for understanding the world's [largest] animal', although difficulties still obtain, such as knowing which animals are making which sounds, as 'the vocalizations are often made by animals that are never seen at all' (Bortolotti 2009: 161, 165).

The melancholy of whale song, in this last sense, is contingent on whales not being seen; on their being heard but not necessarily deciphered; and, in some more extreme cases, such as the low-frequency infrasonic rumblings of the blue whale, on their not being able to be heard (at least by human ears) at all. Its plaintive call, which is capable of covering hundreds, potentially even thousands of miles without a corresponding whale to answer it, has helped carve out the popular view of the blue whale as an 'image of loneliness' (Bortolotti 2009: 89) in the deep ocean, a melancholy effect reinforced by the perception of temporal as well as spatial distance, for whales are ancient beings that both bridge and show the gulf between distinct and incommensurable worlds. Further melancholy effects are produced by the attribution of mystical or religious dimensions to creatures that are clearly of this world, but are also seen as messengers between worlds, as in lofty New Age pronouncements on 'telepathic' whales and 'angelic' dolphins; and in the association of these creatures with water, itself a melancholic element: dream-bound, meditation-oriented, alternately enticing and alienating, as in the further New Age correlation of the deep oceans with outer space.

While these melancholy effects arguably accrue to all whales, it is the humpback that is most closely associated with song, and Rothenberg accordingly sees humpback song as a kind of musical language, refractory to human understanding but perhaps still analysable in musical terms. Humpback songs, Rothenberg suggests, are never the same, continuously changing during the breeding season, which strongly suggests – though this is still far from proven – that their main purpose may be to stimulate female receptivity, making them, as the Harvard-based zoologist E.O. Wilson puts it more confidently, 'the most elaborate single [mating] display known in any animal species' in the world (Wilson, quoted in Whitehead and Rendell 2015: 83). While the motivations for song are unclear, the structure of the songs themselves has been subject to rigorous scientific study. Humpback song, Rothenberg states, is a 'true composition' (11) consisting of patterned and interconnected sounds that are regular, rhythmic, and above all aesthetically satisfying to listen to, offering at once a 'wide frequency range' and a 'variability in harmonic structure' that are intensely beautiful in their own ethereal terms (146). As the marine biologists Hal Whitehead and Luke Rendell describe them, songs are typically arranged in cycles containing about eight 'themes', with each theme consisting of a patterned series of nearly identical 'phrases' (Whitehead and Rendell 2015: 77). However, while songs follow general rules, they tend to be characterized by a wide variety of possible permutations, featuring both 'different content and distinct trajectories of change' (2015: 79).

The sophistication of whale song has been used by scientists to support the increasingly accepted view that whales have culture, and to open up fresh debates on the moral standing of cetaceans, which many now see as being entitled to the personhood that human beings tend to reserve for themselves (see, for example, Whitehead and Rendell 2015; also White 2007 and Chapter 2 of this book). Carrying these debates forward, Rothenberg asks whether whales have an *aesthetic* sense, and whether whale song may be less an instrument of information than a register of aesthetic 'rightness' connected, in turn, with the ways in which 'evolution produces a sense of "rightness", from each species' perspective, about how animals should look and how they should behave' (2008: 149). This squares with Rothenberg's own view of music as a vehicle of affect, an 'emotive repertoire of sounds' (158) which he uses, via the work of the archaeologist Steven Mithen, to suggest that 'music meant much more to our ancestors than it could ever mean to us – as music might mean more to whales than we can ever know' (159). Whale song, for Rothenberg, thus emerges as a kind of ambient music, an emotional support mechanism that whales use to navigate their way through what he melodramatically calls 'a dark, lightless, foggy world' (190). Unsurprisingly, Rothenberg picks out the blue whale as a denizen of this world, seeing the repertoire of low sounds that it emits, which carry vast distances across the ocean, as adding its own musical arrangements to what in effect, and in contrast to Cousteau's now-discredited 'silent world' paradigm, is a vast sea of sound (205).[5] Blue whale sounds carry the distances they do, Rothenberg suggests, because at lower frequencies there is a relative lack of 'sonic interruption' (205). However, the increasing kinds and amounts of sound produced by humans and the machines they create contribute to a very different sound world in which whales and other ocean-going animals find it difficult to communicate. Rothenberg duly cites the American oceanographer Peter Tyack, who states that:

> The deepest whale songs used to carry hundreds of kilometers. Now it turns out that the thing that dominates the frequency range down there is shipping noise. This is where we put the most acoustic energy in the ocean. If that increased low-frequency din is preventing a male and female from finding each other, it is important. It's not as visible as a dead whale on a beach, but it might be much more prevalent. (Quoted in Rothenberg 2008: 209)

Rothenberg echoes Tyack's pragmatic view that more acoustic research is required, but what arguably interests him most is that whales are affective animals capable of inspiring us even if the technology we use to understand them creates an illusion of direct access to animal worlds which remain essentially separate from us. This argument is carried forward by Donna Haraway, who suggests that the flip side of the excitement produced by technologically mediated encounters with non-human animals is regret—that is, the worlds they inhabit aren't actually accessible to us—and by Akira Lippit, who sees our separation from non-human animals as being *represented* technologically, in spectral (e.g. telekinetic) form (Haraway 2008; Lippit 2010; see also Berger 2009).[6] As I have been arguing in this chapter, melancholy is the affect most obviously created by our multisensory encounters with whales, not just in the sense that whales remind us of a multifaceted history of loss, but also in the sense that they come to *embody* these losses: in Pensky's terms, they are the melancholy objects that melancholy subjects must produce in order to validate themselves (1993: 15–16).

In Rothenberg's case, this melancholy 'way of seeing' is routed through what might best be described as a New Age environmentalist ethic that posits the urgent need for greater understanding and communication between human and non-human species while remaining aware of the barriers that continue to stand in the way of such understanding and communication – barriers that cannot be overcome by empathy alone (Heelas 1996; see also Huggan 2013).[7] Such New Age moments are replicated in Rothenberg's text by nostalgic references to the Green idealism of the 1970s: 'No one imagined great whales could make such great sounds, and this was an age when new experience was especially treasured, sought out, and blended with a rush of sensory possibilities. We dreamed of a better, more joyful world, and singing whales would be part of it' (Rothenberg 2008: 19). Rothenberg's own jazzy experiments in 'interspecies music' (75) hark back to this now largely bygone age, one witheringly described by the historian of science D. Graham Burnett (2012: 599) as an era of 'peaceniks, freaks, and ecoterrorists' in which, in the environmental anthropologist Arne Kalland's equally dismissive language, caring for whales came to be a 'metaphor for being good' and whale society a complementary 'metaphor for the lost human paradise or utopian world' (Kalland 2012: 41).

To be fair to Rothenberg, *Thousand Mile Song* is too playfully eccentric to fully affiliate itself with the New Age or any other quasi-spiritual movement, while at times the text is notably disabused of its own New Age predilections, for example in the epilogue when Rothenberg joins other pleasure-seeking tourists on a boat off the Hawai'ian coast, communing mainly with each other while claiming to be celebrating the 'passion and beauty of [the] giant sirens of the ocean depths' (2008: 221). The following passage sums up the hedonistic atmosphere perfectly:

> 'Come Join the Spirit Sailing Journey to the Cetacean Nation,' says the e-mail invite, full of tiny print listing all the fabulous love and music that will be on board. 'Leave your shoes on the beach, swim out to our catamaran, drums and shakers will be on board. Our guest musicians will play cellos, harps, and guitars. We'll have flippers, you can dive in with our giant friends.' As soon as we leave the harbor, clothes tend to come off. A woman called Mahana, who was once called Lauren, waves and shouts the old Hawaiian word for humpback out to the sea. '*Kohola, koholaa,* we love you, come close to us, we're waiting.' The passengers start to sing and dance, the warm sun beats down on us, and we see the first great humpbacks cavorting in the bay. It's prime mating and calving season, and they're everywhere. A few hundred yards offshore and we're mostly naked, trying to tune in to that great whale energy. (222)

Rothenberg's positioning of himself as a tourist rather than a scientist, albeit one who is sceptical of the antics of would-be cetacean tourists and the opportunism of 'rogue whale watchers' (235) who will do whatever it takes to get close to whales, offsets his own maverick efforts to make music with them – itself an opportunistic pursuit if also one tinged with a sobering awareness that the recent history of human reception of whale song reveals a story of beauty under pressure, melancholic in both its acoustic intensity and its conservationist intent. Notwithstanding, whale watching – in all its myriad forms – represents the best chance most people have to see whales in their natural state, if few among them have the further opportunity of listening to them underwater. As Kalland says, prior to the exponential growth of the whale-watching industry, now a two-billion-dollar global concern, very few people indeed had experienced direct engagement with cetaceans. Instead, most of us met whales 'only through their representations, and our knowledge about [them was] mostly built on representations presented through books, photos and movies' (2012: 28). If this still remains the case for many today, whale watching has become part of the reality of experiencing whales for millions of people. But how 'real' is this experience, and what emotions does it bring to the fore? Is it merely a pleasurable pursuit, or are there underlying motives? And, in the still incomplete transition from whale hunting to whale watching, is there a melancholic residue – a buried understanding that yesterday's losses may not necessarily translate into today's gains? These are some of the questions that the next section of this chapter seeks to address, taking its cue from the Scottish poet and nature writer Kathleen Jamie's only partly tongue-in-cheek observation that whale watching 'proceeds like a kind of theology – by glimpses, sightings, a dorsal fin, a rolling back'; and that, as such, it is best understood as 'a pursuit for the regretful; all might-have-beens and what-did-we-miss?' (2005: 145).

Whale/watch

Since its modest beginnings in 1950s' California, whale watching has evolved into one of the most significant and commercially lucrative forms of contemporary global marine ecotourism – ostensibly a conservation-oriented mode of tourism which, in the Caters' definition, 'takes place in coastal and marine settings; is designed to help conserve the local environment while also benefiting local communities; and aims to educate visitors about this environment while also motivating them to re-examine how their activities impact upon the Earth' (Cater and Cater 2007: 9; see also Garrod and Wilson 2003). Recent research on whale watching, however, has complicated the too-easy view that it represents an authentic alternative to watching animals in captivity; and that, motivated as it apparently is by conservationist imperatives, it marks the transition from a brazenly exploitative age of industrial-capitalist extraction, embodied in the destructive practices of whale hunting, to a more sustainable era in which education trumps exploitation, and the safety and well-being of the whales themselves are of paramount concern. A good recent example here is James Higham, Lars Bejder, and Rob Williams's state-of-the-art 2014 study on the social practices and cultural politics of whale watching, which marshals substantial evidence that whale watching is by no means always educationally sound or environmentally friendly, while the anthropologist Adrian Peace claims that it is 'as emblematic of mankind's early twenty-first century hubris as whale slaughter was expressive of its arrogance in the twentieth century', not least by taking it for granted that 'the worth of whales is to be determined by our consumption-fixated world-view alone' (Peace 2005: 206–207).

Still, it would seem as unwise to dismiss whale watching's conservationist potential as it would be to assume it; similarly, sustainable ecotourism should neither be considered as a naïve tautology nor as a flagrant contradiction in terms. More research is needed, Higham et al. insist, on how and to what extent whale watching can be harnessed to local community understandings as well as to broader conservationist initiatives, while more is also required on how and to what extent different individuals and groups of people interact with their environment, as well as on how they interact with whales (Higham, Bejder, and Williams 2014; see also Corkeron 2004). While acknowledging the need for updated empirical work of this kind, my own approach in this chapter (as elsewhere in this book) is somewhat different: more philosophically inclined if also more historically inscribed. Broadly speaking, the argument I want to make here is that whale watching, while clearly motivated in different ways that are influenced by a variety of social and cultural factors, is a largely *compensatory* activity in which the pleasure gained from looking at animals is counterbalanced by the melancholic awareness of their loss or disappearance. This argument follows John Berger's view – expressed in his famous 1980 essay 'Why Look at Animals?' – that when we look at animals we do so in order (1) to confirm our own presence in face of them, and (2) to compensate for their loss. Berger's essay is paradigmatically melancholic in its mournful view that animals have mostly disappeared from our lives, and that as a result we have become a quintessentially lonely species, confirmed in our loneliness by the modern capitalist system that governs our relations to other people, animals, and things (Berger 2009 [1980]: 11). What Berger means by this is not that animals have *actually* disappeared (although many species certainly have), but that they have become supplementary rather than integral to our existence (11). Modifying this slightly, I would suggest that they are still integral, but that many of us have chosen to forget this, just as most of us are increasingly blind to the wildlife which – in some cases in alarmingly diminishing numbers – continues to surround us, and upon which we depend far more than we might imagine in our daily lives (Louv 2010).[8]

Berger's work remains influential despite the criticism it has received for privileging the visual (for human encounters with animals are nothing if not a multisensory experience); for lapsing into the nostalgic (an effect of the fundamentally melancholic sensibility that drives it); and, perhaps most tellingly, for not answering the question it poses. For, as Garry Marvin among others suggests, it isn't clear from the essay what actually motivates people to look at animals even though it *is* clear that the global capitalist system into which they are incorporated tends to reduce animals themselves to commodities within a symbolic economy of spectacle that converts 'their' bodies into 'our' images, and 'their' real or imagined disappearances into 'our' narcissistic feelings of loss (see Marvin 2005; also Baker 2001, Lippit 2010). While I agree with all of these criticisms, I would add that the larger question of why we look at animals needs to be seen alongside the narrower issue of why (also how) we look at *particular* animals, for as Marvin argues, albeit stating the obvious, 'a peasant does not look at his pig in the same way that a zoo visitor looks at a lion, nor does the hunter look at a deer in the same way that a companion animal guardian looks at his dog or cat' (2005: 4).

This lack of species differentiation can be seen even in the more sophisticated work that is a product of the so-called 'animal turn',[9] at least some strands of which make the category mistake of seeing all animals as companionate creatures, or which offer sweeping, unwittingly sentimental distinctions between an *objectifying* gaze that confirms our power and imagined superiority over animals and a *reciprocating* gaze that acknowledges our kinship with them in what eco-philosophers and animal theorists have increasingly taken to calling a 'multispecies' world (Woodward 2001: 3; see also Haraway 2008, Kirksey 2014). The idea of the reciprocating gaze tends to assume that animals are as interested in us as we are in them, and that all animal species – humans included – form part of what Donna Haraway calls a grand planetary 'subject- and object-shaping dance of encounters' in which 'actual encounters are what make [lively] beings' in the first place, and a technologically enhanced mixture of 'commerce and consciousness, evolution and bioengineering, and ethics and utilities' comes into play (Haraway 2008: 4, 67, 46). How 'actual', however, are our encounters with other animals? This depends of course on linked sets of particularities (particular conditions, particular humans, particular animals). But it also depends, as Haraway recognizes, on multiple levels of mediation in which the animals we *look at*, which are increasingly shaped by modern technologies as well by longer histories of symbolic correspondence and association, are simultaneously *looked for* in terms of what we think, and what others have persuaded us into thinking, they represent (Haraway 2008; see also Baker 2001).

This entanglement of human/animal bodies and technology is most evident in the history of nature documentary and wildlife film. This history brings together *the look* (how we see or would like to see ourselves) and *the gaze* (how we are positioned socially and culturally). These in turn constitute a visual regime, apprehended primarily via the camera, in which the play of human/animal identities is framed and fashioned through 'an irreducibly exterior image which intervenes between it and the subject', and where even the most reciprocally oriented of human/animal imaginaries is filtered back to the subject via the controlling mechanisms of an objectifying gaze (Silverman 1996: 135; see also Brower 2011, Burt 2002, Woodward 2001). I would argue, however, that these mechanisms are also at work in more ostensibly 'direct' encounters with animals, such as those experienced in whale watching. Whale watching, after all, is nothing if not a staged spectacle in which our reactions to the whales we (wish to) see are partly conditioned by the second-hand knowledge we already have about them. These reactions are then further shaped and influenced by the suspenseful atmosphere created around them, which often depends on the carefully manufactured production of authenticity and on a technology-enabled, performance-oriented interpretation of cetacean behaviour which, if not necessarily inaccurate, is designed to please (Kalland 2012; Peace 2005).

Commercial whale watching is mostly family entertainment these days, and while it has the capacity to generate considerable excitement among young and old alike, its feature elements of surprise and anticipation, while these can never be monitored completely, are carefully rationed out. There is thus a continuum from the idea of *whales on demand*, which are only as far away as the next television screen, to the idea of *whales on display*, where captive animals are made available to be looked at, to the freer idea of *cetacean encounters*, where whale sightings are subject to the law of probabilities, but sightings are still probable, and there is less emphasis on orchestrated performance, but there is performance nonetheless. All the evidence suggests that whale watching is a 'high-satisfaction' industry (Higham, Bejder, and Williams 2014; Orams 2000). Still, the possibility of disappointment is built in to whale-watching tours. In this and other respects, these tours match the structural conditions that obtain within the mass-tourism industry, which is designed in part to frustrate the expectations of its paying customers, just as it offers opportunities to sample new experiences that brush off old disappointments, feeding off regret, possibly even recrimination, but providing (for a fee) the chance to start all over again (MacCannell 1976).

It is still worth asking what it is about *whale watching*, in particular, that creates a disappointment that goes well beyond whether one tour or another lives up to its expectations, and that has relatively little to do with specific experiences, for example whether whales 'in the wild' are actually seen or not. Here it might be useful to come back to Jamie's aforementioned claim that modern-day whale watching 'proceeds like a theology' (2005: 145). Jamie is referring, I think, not so much to the arcane notion of hidden gods or the faith-bound possibility of our transcending our earthly existences, but to what I would see as being a *dialectics of appearance and disappearance* that is integral to human–cetacean encounters – one in which we are continually made aware of our own lack of vision, and in which it is ultimately the whale rather than the viewer that decides when, where, and how much to display. It is not for nothing that the iconic photographic image of the whale is that of the flukes, the last trace of a creature that is about to escape, possibly never to be seen again, from our restricted field of vision (Brower 2011; see also cover image). Similarly, sightings of whole whales – occasional jumps and breaches notwithstanding – are relatively rare in whale-watching tours; it is much more usual to see parts of animals that allow for, and indeed stimulate, an *imagination* of the whole. Such tours remind us, in other words – as do whales themselves – of the singular limitations of our vision. These limitations also pertain to our knowledge and understanding of creatures that have become 'ethically, spiritually, and ecologically important to us' (Stewart 1995: 13), but that even experts admit are much more talked about than seen, and much more readily imagined than scientifically known.

My own experience of watching whales is similarly limited. I haven't watched an orca display and probably wouldn't own up if I had, but I *have* seen orcas twice in their natural state, once – wholly unexpected, and thrillingly close – off the coast of Salt Spring Island in south-west Canada, and once on an official whale-watching tour in the same region, departing from Vancouver and threading its way through the American-owned San Juan Islands, across the Canadian–US border to the city's south. The following is a brief narrative distillation of that tour, which is largely reliant on memory, as when the tour was taken I had no intention of using it in my academic work.

> *It's a beautiful sunny day and my companion and I are careful to take hats and water; as I discover once we've left the harbour, it's just as well I brought a rain jacket as well, as though there's no trace of rain, the wind soon gets up. I have a camera of course, but old-fashioned technophobe that I am, I'm ambivalent about using it; it seems a shame to agitate about capturing things on camera when there's so much to experience first-hand. There are about a dozen of us on board: all ages, all shapes and sizes, all nationalities. It's a good-humoured bunch on the whole, though a few openly complain about the noise of the boat's engine, while others – myself included – wonder what effect this is likely to have on the whales.*
>
> *After about an hour at sea with not a whale in sight, things get uneasy. One Spanish woman, who has been intermittently sick since the start, has since gone to sleep inside the boat and seems no longer inclined to participate, while a young British boy, quickly hushed by his embarrassed-looking parents, says he now wishes he hadn't come and why is everyone wasting his time. The boy peers angrily at his phone instead; most of the rest of us are anxiously scanning the water. The scenery is great, and it strikes me that the pleasures of whale watching aren't necessarily tied to whales at all, but rather to the experience of being out on the water. Then I correct myself: I'm getting cross too that I haven't seen a whale.*
>
> *I needn't have worried: a pod of orcas has just been spotted further on up the bay, and the boat accelerates towards it. As I've heard is frequently the case, especially in high season, other commercial boats have been given the same information, and one in particular seems destined to become our rival, first tailing in our wake then cheekily moving past. The anxiety mounts: what if this boat gets there first, and what if they get a better view? What if we miss out entirely? The anxieties are allayed somewhat by the first sighting: three whales – two adults and a calf – making serene progress in the coastal waters of the bay. There's a certain jostling for prime viewing position in the boat; one small German group has planned ahead by shoehorning themselves into the available space on the upper deck, which is relatively unobstructed, but soon there are pointing cameras everywhere. Mayhem ensues when the*

whales suddenly change direction, speeding up as they do so. The boat clumsily manoeuvres round to tag them, and eventually catches up. Alas, the whales seem more interested in our rival boat than ours: they seem to be cavorting around it. The Germans are miffed; one large American tourist curses. Again we needn't have worried though: the whales are starting to move back in our direction; and before we know it, thrillingly, all three have passed right under the boat. They move effortlessly, their black-and-white backs sliding in and out of the water, sparkling in the sun. I am taken aback by their size, though they certainly appear to be placid. Killer whales are known for their ferocity, but they rarely attack boats and are mostly indifferent to human presence. This seems to be the case here, as the pod plays cat and mouse with the two skirmishing tourist boats that inelegantly trail it. Our boat seems to have the upper hand, but then the whales veer off again, seeming more agitated. They're on the hunt, our captain cheerfully says, and suddenly the waters are churning; we get partial views of diving and what looks like a jump, but damn it the main scene of the action is closer to the other boat. The Germans are miffed again, but the excitement is largely shared between us. Eventually the other boat moves off and we move in, but it seems that feeding time is over; the whales are on the move again, and quickly, heading out to sea.

One last glimpse or two and they're gone, but our time is up anyway, and the journey back in is as uneventful as the journey out. I manage to clamber onto the upper deck, where I sunbathe; my companion, who is Peruvian and doesn't like the sun much, has moved inside the boat in search of shade. An hour later, we're back on land, trying to remember where we parked the car and hoping we're not going to get a ticket; Canada is a country of permits and licences, though I'm grateful that its whale-watching industry is also reasonably well controlled. Maybe I should have been worried though: just a few weeks after our trip, there's a horrendous accident off the west coast of Vancouver Island when a whale-watching boat crashes into the rocks in bad weather and several people are killed. Apparently the boat got too close in tracking a pod of whales and was caught by a rogue wave, though other, more openly accusatory interpretations of events are offered. I'm reminded that whale watching, though largely safe, is an exercise in risk management, both for the people who participate in it and, of course, for the whales.

Recounting these events reminds me why I'm neither a creative writer nor a social anthropologist: I lack the descriptive and narrative skills to be the first and the facility with people to be the second. Still, it also confirms to me that whale watching is largely an anticipatory activity that puts patience to the test, although certainly no more so than other marine leisure activities – fishing for example – which, like whale watching, are probably more about soaking up atmosphere and charting changing mood patterns than they are about the human–animal encounter itself. I'm struck too that 'melancholy' is perhaps the wrong term to use after all, although whale watching, like other supposed relaxation pursuits, is paradoxically anxiety-inducing, motivated less by the painful awareness of past and present loss than by the more selfish fear of missing something: of losing out. Theorizing whales is always likely to be challenged when encountering them in practice; perhaps all the more so when those encounters are simultaneously recognized as being elaborately staged. Still, at some level the experience of watching whales gives us pause for thought on how we thirst for the authenticity of natural things even as we acknowledge the gap between what we see and what we want to see, and even though we recognize that there is no such thing as unmediated access to the natural world (Brower 2011: 86, 197). Melancholy may yet describe this gap and the conflicting emotions which surround it. Meanwhile, as previously argued, the trope of the 'melancholy whale' operates as a symbolic figure, not just for a circumscribed existence but for the effective incommensurability of worlds that we humans clearly influence, but that we struggle – wilfully at times, involuntarily at others – to comprehend. The final section of this chapter looks more closely at the composition and meaning of these 'worlds', assessing to what extent whales, and other animals, are encircled within separate realms of their own making, or whether these realms are drawn together, each bounded in the other, and each part of a larger network of relations that continually composes and recomposes itself (Buchanan 2008: 161).

Whale/world

As I have been arguing throughout this book, whales are subject to a wide variety of different 'ways of seeing' in which our understanding of them is inseparable from our knowledge of their cultural representations, and in which these representations are both constructions of a wider culture and work towards constructing that culture themselves (Baker 2001: 25; see also Fudge

2005). To put this in another way, whales belong to a 'society of the spectacle' (Debord 1994) characterized by the density of images that surround us, and by the conversion of these images into commodities that function within a global economy of symbolic exchange. This implies that, however much they are predicated on experiencing the real, human–cetacean encounters are always mediated to some degree, and that a full account of these encounters requires *both* updated empirical work on whale watching, for which the most reliable methods are those provided by tourism anthropology, *and* a broad-based historical understanding of how whales have been made subject to a symbolic politics the far-reaching effects and consequences of which can still be seen today.

This suggests in turn that whales are, as Donna Haraway (2008) might put it, 'material-semiotic' entities, both living creatures attached to material histories and bundles of multivalent – frequently allegorized – signs. In this final section of the chapter, I want to explore the material-semiotic status of whales one last time by associating it with different, multiply interconnected *worlds*. These include the world of *empire*, which, as previously argued, is closely interwoven with the history of whaling (see Preface; also Chapter 1 and Chapter 2); the mutually informing worlds of *global commerce* and *industrial capitalism*; and the world of *human–animal relations*, itself a material-semiotic entity in which animal bodies, exploitable for commercial purposes, are also rendered 'symbolically available' as multifunctional signs (Baker 2001: 5). Nicole Shukin puts this well when she suggests that animals have both material and metaphorical currency, and are thus manipulated in the interests of capital in both a literal and a figurative sense (2009: 5). In the case of the whale, one of the common denominators for both forms of manipulation is *violence*. Hence the particular relevance of her term 'rendering', which, in bringing together the 'carnal presence' of the animal and the various cultural meanings that are mapped onto it, also joins the violence done to animals to the violence of representation itself (2009: 23). Nor is this violence consigned to the past; for even the protected New Age whale is subject to other renderings, whether it functions materially/symbolically as museum exhibit or as aquarium display item. There are other assailants, too, such as marine pollution, climate change, and not least international political intrigue; for the goal of protecting whales plays out in a fraught cultural–political arena in which the global currency of cetaceans as conservationist icons is manipulated at national level, with some countries vying to be seen as more 'environmentally friendly' than others – with much more than just national pride at stake (Burnett 2012; see also Chapter 1).

Another common denominator, I have been suggesting more closely in this chapter, is *melancholy*. The figure of the melancholy whale is linked to a history of loss that, like melancholy itself, becomes inexorably personalized; or, perhaps better, it is linked to a compound history of external losses – some real, others imagined – that become destructively embedded within the individual self. The spectrality of the whale plays an important role here. For Philip Hoare, the trademark elusiveness of the whale is tied in with memories of loss, but also with premonitions of future disappearance; as he says after going on a whale-watching tour, 'It was hard to look on these huge creatures and think of a time when they might not be there' (2009: 214). Here as elsewhere in his work, Hoare has difficulty controlling emotions that are influenced, to some degree, by a 'cetacean imaginary' characterized by myth-saturated spectral and/or apocalyptic visions: whales as apparitions, surfacing momentarily only to return to the deep ocean; whales as fated creatures, destined to die out (2009: 214; see also Chapter 1). More theoretically minded critics like Shukin and Lippit interpret the spectrality of the animal more generally by pointing out that animal signs function simultaneously as *fetishes*. Although neither mentions whales specifically, Shukin's analysis of the fetish is particularly relevant to both the 'real' and 'represented' histories of human encounters with whales. Animal signs, she says,

> function fetishistically in both Marxian and psychoanalytic senses: that is, they endow the historical products of social labor to which they are articulated with an appearance of innate, spontaneous being, and they serve as powerful substitutes or 'partial objects' filling in for a lost object of desire or imaginary wholeness that never did or can exist, save phantasmatically. (2009: 3; see also Lippit 2010)

Extrapolating from this, we might say that whales are both *commodity* fetishes in the Marxian sense that the history of their commercial use hides the exploitative labour (as well as the considerable violence) that produces it, and *nostalgic* fetishes in the psychoanalytic sense that they operate as substitutes for a lost 'wholeness' that never existed in the first place. Needless to say, both functions are deeply anthropocentric. In the first case, whales are literally rendered down for human use (as lighting, soap, margarine, etc.); in the second, they are figuratively deployed as 'powerful metaphors for a world that modern human beings [have] allegedly lost' (Kalland 2012: 29). The second function, in particular, reminds us why whales not only bring together different worlds but also become metonymic stand-ins for the world itself. Again, there are both literal and figurative understandings of this. On a literal level, whales are rarely seen whole; rather, their wholeness must be imagined. On a figurative level, the invisible or only partly visible body of the whale is rhetorically inflated through cetacean iconography, such that whales – already huge – become even bigger than themselves, expanding in the imagination to become worlds of their own, or even to fill the entire world itself (Hoare 2009).

It seems worth asking what these worlds are, which requires a brief excursus into ethology. The question of what constitutes an animal world is usually traced back to the early twentieth-century ethological work of the Estonian scientist Jakob von Uexküll, for whom each and every animal effectively creates and inhabits its own experiential world (*Umwelt*). This relates to von Uexküll's phenomenological view – usefully glossed by the philosopher Brett Buchanan – that 'there is no objective reality in the form of objects, things, or [indeed] the world; there is nothing outside of the individually subjective experiences that create a world as meaningful' (Buchanan 2008: 13). The crucial question, for von Uexküll, is to what extent animals are contained within their own worlds and are able to appreciate and interact with other worlds beyond them. Buchanan takes up this question in relation to twentieth-century Western philosophy, scanning Heidegger's, Merleau-Ponty's and, finally, Deleuze's work for the insight it provides into both the observable study of animal behaviour (ethology) and more notional understandings of the relationship between humans and other animals. In brief, Buchanan sees Heidegger as confirming von Uexküll's view that animals remain largely 'immured within their own captivity' (Buchanan 2008: 113). Deleuze, on the other hand, gestures towards an alternative understanding of the human–animal relation in which the connection between all living beings (whether designated as 'human' or 'animal') is measured in terms of their mutual capacity to affect one another, making ethology less a study of animal behaviour than a kind of affective registration of 'the continual becomings that compose different bodies' and of the different ways in which human/animal bodies undergo transitions, continually creating and recreating themselves (Buchanan 2008: 158–159). As Buchanan notes, Deleuze's idea of a mesh or network assembled through different 'rhizomatic' conjunctions and associations (170), within which 'everything is in a process of becoming' (172), is a long way from von Uexküll's circumscribed, soap-bubble-like animal worlds. Instead, Buchanan says, the 'soap bubble has burst, with various repercussions' (175), not least among them the transition from what he calls 'a biology of the organism' to 'an ethology of affects' in which continual process and movement belie the idea of a single, achieved state (182). Here, Buchanan himself traces a transition from von Uexküll's view that 'there are as many worlds as there are living beings' (187) to Deleuze's post-Heideggerean conviction that 'being-in-the-world' is less important than 'the becoming of the world' (189).

More recently, eco-philosophers such as Timothy Morton have gone further still to shatter the very idea of 'world' as a contained space with definable boundaries. For Morton, 'World is more or less a container in which objectified things float or stand' (2013: 99); it is also a human-centred concept designed to convey the illusion of human mastery and control. We would do well, Morton suggests, to let go of 'world' and, in so doing, develop a greater sense of our entanglement with others, which might in turn guide our ethics. 'World', says Morton in his characteristically breezy way, is no more than a blurry 'aesthetic effect' based on a false sense of distance; it is neither more nor less than a framing device, a 'world picture' that we could all do without (2013: 104, 106).

While there is scientific support for Morton's Deleuzian vision of the porosity of existence on both a planetary and, indeed, a cosmic scale, I find his views too extreme. This applies as well to his earlier work on 'nature', which, much like 'world', is prematurely dismissed as an objectifying concept that separates humans from the other material agents, both organic and non-organic, with which they continually interact (Morton 2007).[10] Both 'world' and 'nature', I would argue instead, retain considerable rhetorical if not descriptive power as discursive tools for environmentalist calls to action, which are not particularly well served by melancholy thought. Morton's thought, I would argue further, eventually succumbs to its own melancholy inclinations: hence his wavering – probably most evident in his recent work on 'dark ecology' (Morton 2016) –

between utopian visions of planetary flourishing and an involuntary fatalism in which such visions, if not recognized as meaningless, are impregnated with despair.

It seems only appropriate then that, in his influential 2013 book *Hyperobjects*, Morton makes several references to whales, using the parable of Jonah and the Whale in particular to make the case that the whale that Jonah is trapped inside is itself a hyperobject, 'a higher-dimensional being than ourselves' (2013: 133). Global warming, the ostensible subject of Morton's book, is likened to being inside the belly of the whale: in other words, it is a grand-scale, potentially intractable problem, 'an ecological emergency inside of which we have now woken up' (2013: 201). In the parable itself, Jonah is eventually saved; Morton gives us little indication that fortune will again favour us. At the same time, Morton's is the last example in this book of the double-edged nature of cetacean allegoresis, which I have been associating in this chapter with melancholy. Over millennia, cetacean allegories have served a wide variety of different purposes, by no means all of them debilitating or destructive, and by no means all of them, to recall Stamelman's earlier assertion, associated with the degenerative languages that 'death speaks or writes' (quoted in Pick 2011: 76). However, such allegories, which revolve obsessively around the figure of the whale, tend to be 'sicklied o'er with the pale cast of thought'. And, as Hamlet once similarly recognized, clouds may be 'very like' whales, but they are emphatically not whales – reminding us that the melancholy condition, in confronting loss, is also adept at displacing that loss onto something else.

Notes

1 Shukin defines 'rendering' in terms of a *double entendre*, signifying both 'the mimetic act of making a copy, that is, reproducing or interpreting an object in linguistic, painterly, musical, filmic or other media', and 'the industrial boiling down and recycling of animal remains' (2009: 20). Although Shukin uses this dual function to show the general representational and economic workings of what she calls 'animal capital', its more specific relevance to whales should be obvious here.

2 There is no hard-and-fast way of differentiating between affects and feelings or emotions, although one view, closely associated with Deleuze's philosophy, is that

affects refer less to recognizable emotions, which are socially structured, than to more amorphous precognitive states. Perhaps the best distinction I've come across is in Sissy Helff's essay in Jopi Nyman and Nora Schuurman's 2016 collection *Affect, Space and Animals*. As Helff explains in distinctly Deleuzian terms, 'Where *feeling* is often used to refer to phenomenological or subjective experiences, *affect* is often taken to refer to a force or intensity that can belie the movement of the subject who is always in a process of becoming. […] Affects do not refer to a "thing" or substance, but rather [to] the processes that produce bodies as always open to others, human and nonhuman, and as unfinished rather than stable entities. […] Emotions, in contrast, are those patterned brain/body responses that are culturally recognisable and provide some unity, stability and coherence to the felt dimensions of our relational encounters' (Helff, in Nyman and Schuurman 2016: 97).

3 Thanks to Kate Marshall for pointing out the term 'pseudo-extinction' to me.

4 In her book *Frames of War*, Judith Butler distinguishes between 'grievable' and 'ungrievable' lives in the context of war, where the former concept refers to lives considered worth mourning because they function in defence of some collective idea, for example the community or the nation, whereas the latter refers to lives not considered worth mourning because they are not considered to be lives at all. Extending Butler's work to the animal domain, Anat Pick suggests that the idea of vulnerability embedded within 'grievable life' crosses over between human and non-human histories: naming this idea, she speaks of the 'creaturely', which she sees in terms of the vulnerable body which, whether human or animal, serves as an example of 'bare life' – 'bare in the sense of being susceptible to the interventions of power' (Pick 2011: 15). Whales clearly belong to the category of 'grievable life', which is often expressed in melancholy terms, and even the most avid of whale hunters – like hunters in general – tend to mourn the very animals they (wish to) destroy. Consider Melville's Ahab, a man consumed by rage but also haunted by a kind of timeless remorse: '"Swim away from me, do ye?" murmured Ahab, gazing over into the water. There seemed but little in the words, but the tone conveyed more of deep helpless sadness than the insane old man had ever before evinced.'

5 Jacques-Yves Cousteau's book *The Silent World*, first published in 1953, is probably best known today for the award-winning documentary of the same name, which was released three years later. The idea of the subaquatic realm as a 'silent world', which would become further popularized with Cousteau's numerous films and television series, has since been comprehensively overturned by marine biologists and oceanographers, whose acoustic research has revealed the same world to be positively garrulous with all manner of underwater conversations and animate/inanimate sounds.

6 The central thesis of Lippit's Derrida-inspired book *Electric Animal*, originally published in 2000, is that – in his own words – 'Modernity can be defined by the disappearance of wildlife from humanity's habitat and by the reappearance of the same in humanity's reflection of itself: in philosophy, psychoanalysis, and technological media such as the telephone, film, and radio' (Lippit 2010: 2–3; see also Chapter 1). This leads him, in turn, to explore the figure of the 'spectral animal', which he sees as having become part of a technological sphere that exists as 'the afterlife of language' (161), and to liken animals to films as 'project[ing] from a place that is not a place, a world that is not a world' (195). As previously argued (see Chapter 1), it is one thing to say that the *idea* of the animal disappears, but another to say that animals *physically* vanish. Shukin probably says it best, warning that figures of animal spectrality risk 'draining historical materiality out of the sign of animal life' (2009: 36).

7 See my 2013 book *Nature's Saviours* for a detailed examination of New Age environmentalism, which (as the ironic title of the book implies) often draws on the rhetoric of salvation to confer a spiritual dimension on earth-bound conservationist work. An uncritical use of empathy to cross human and animal divides is a feature of New Age environmentalist thought, which, despite the undoubted silliness around its fringes, is by no means as frivolous or wrong-headed as it is often made out to be. For balanced analyses of New Age philosophies and world views, see Heelas 1996 and Ross 1991.

8 An interesting variant of this is the American journalist Richard Louv's concept of 'nature-deficit disorder', a condition with varying physiological and psychological symptoms that he attributes to the increasing lack of outdoor activity of today's adults and children alike. While Louv's views, encapsulated in his book *The Last Child in the Woods* (a melancholic title if ever there was one), are unsubstantiated at best and at worst exemplify the kind of cod psychology that continues to hover at the edges of the Green movement, they are useful nonetheless in giving a more popular flavour to Berger's more sophisticated but, in its own way, no less sentimental work.

9 The 'animal turn' is the general descriptive term given to a recent cross-disciplinary movement, closely associated with the work of 'posthumanist' thinkers such as Cary Wolfe and Donna Haraway, which encompasses broad questions of signification and communication (including the representation of animals in literature and the media); new considerations of ethology, ethics, and epistemology; and a radical rethinking of the human–animal relationship (including a critical interrogation of the different meanings embedded within

the categories of 'human' and 'animal' themselves). For a useful summary of the main components of the 'animal turn', see Weil 2010.

10 While Morton would be unlikely to label himself in this way, his thought is sometimes associated with the 'new materialism'. The 'new materialism' is probably best understood as a loose amalgam of theoretical approaches revolving around the idea that all matter – whether animate or not – possesses agency, and that material objects interact with and imprint themselves on one another in ways that demand a rethinking of social and political relations informed by what Morton calls, rather mysteriously including the definite article, 'the ecological thought' (Morton 2010). While 'new materialists' insist that their work is political to the core, the abstruseness of some of their ideas is not particularly conducive to political action, while the salutary recognition that all matter possesses agency begs the question of how *much* agency, and in relation to *what*. Similarly, while 'new materialists' are quite right to stress the limitations of human control, this begs the further question of how to exercise due responsibility in environmental matters that require a considerable degree of human management, even if 'management' isn't always the answer, and even if there are high-level problems (like climate change) that require urgent attention without necessarily being 'solved'. For useful introductions to some of the main concepts informing 'new materialist' thought, see Bennett 2010; also Coole and Frost 2010.

Works Cited

Armstrong, Philip (2008) *What Animals Mean in the Fiction of Modernity*. London: Routledge.

Baker, Steve (2001) *Picturing the Beast: Animals, Identity, and Representation*. Champaign, IL: University of Illinois Press.

Bennett, Jane (2010) *Vibrant Matter: A Political Ecology of Things*. Durham, NC: Duke University Press.

Berger, John (2008) [1972] *Ways of Seeing*. London: Penguin.

Berger, John (2009) *About Looking*. London: Bloomsbury.

Bortolotti, Dan (2009) *Wild Blue: A Natural History of the World's Largest Animal*. Toronto: Thomas Allen Publishers.

Brower, Matthew (2011) *Developing Animals: Wildlife and Early American Photography*. Minneapolis: University of Minnesota Press.

Brown, Laura (2010) *Homeless Dogs & Melancholy Apes: Humans and Animals in the Modern Literary Imagination*. Ithaca, NY: Cornell University Press.

Buchanan, Brett (2008) *Onto-Ethologies: The Animal Environments of Uexküll, Heidegger, Merleau-Ponty, and Deleuze*. Albany, NY: SUNY Press.

Burnett, D. Graham (2012) *The Sounding of the Whale: Science and Cetaceans in the Twentieth Century*. Chicago: University of Chicago Press.

Burt, Jonathan (2002) *Animals in Film*. London: Reaktion Books.

Butler, Judith (2010) *Frames of War: What Is a Grievable Life?* New York: Verso.

Cater, C. and E. Cater (2007) *Marine Ecotourism: Between the Devil and the Deep Blue Sea*. Wellingford, OX: CAB International.

Chambers, Ross (1987) *The Writing of Melancholy: Modes of Oppression in Early French Modernism*. Trans. M.S. Trouville. Chicago: University of Chicago Press.

Coole, Diana and Samantha Frost (eds) (2010) *New Materialisms: Ontology, Agency, and Politics*. Durham, NC: Duke University Press.

Corkeron, Peter (2004) 'Whale Watching, Iconography, and Marine Conservation'. *Conservation Biology* 18.3: 847–849.

Cousteau, Capt. Jacques-Yves, with Frédéric Dumas (2004) *The Silent World*. Washington, DC: National Geographic Classics.

Debord, Guy (1994) *The Society of the Spectacle*. Trans. D. Nicholson-Smith. New York: Zone Books.

Ellis, Richard (1991) *Men and Whales*. New York: Knopf.

Fudge, Erica (2005) 'The History of Animals'. *Society & Animals* 13.1: 13–31.

Garrod, Brian and Julie Wilson (eds) (2003) *Marine Ecotourism: Issues and Experiences*. Bristol: Channel View Publications.

Giggs, Rebecca (2015) 'Whale Fall'. *Granta* 133: 1–22.

Haraway, Donna (2008) *When Species Meet*. Minneapolis: University of Minnesota Press.

Heelas, Paul (1996) *The New Age Movement*. Oxford: Wiley-Blackwell.

Heise, Ursula (2016) *Imagining Extinction: The Cultural Meanings of Endangered Species*. Chicago: University of Chicago Press.

Higham, James, Lars Bejder, and Rob Williams (eds) (2014) *Whale-watching: Sustainable Tourism and Ecological Management*. Cambridge: Cambridge University Press.

Hoare, Philip (2009) *Leviathan or, the whale*. London: Fourth Estate.

Höijer, Birgitta (2010) 'Emotional Anchoring and Objectification in the Media Reporting on Climate Change'. *Public Understanding of Science* 196: 717–731.

Huggan, Graham (2013) *Nature's Saviours: Celebrity Conservationists in the Television Age*. London: Routledge.

Jamie, Kathleen (2005) *Findings: Essays on the Natural and Unnatural World*. London: Sorts of Books.

Kalland, Arne (2012) *Unveiling the Whale: Discourses on Whales and Whaling*. Oxford: Berghahn Books.

Kirksey, Eben (ed.) (2014) *The Multispecies Salon*. Durham, NC: Duke University Press.

Krasznahorkai, László (1998) *The Melancholy of Resistance*. Trans. G. Szirtes. New York: New Direction Books.

Lippit, Akira Mizuta (2010) *Electric Animal: Toward a Rhetoric of Wildlife*. Minneapolis: University of Minnesota Press.

Lopez, Patricia and Kathryn A. Gillespie (eds) (2015) *Economies of Death: Economic Logics of Killable Life and Grievable Death*. London: Routledge.

Louv, Richard (2010) *The Last Child in the Woods: Saving Our Children from Nature-deficit Disorder*. New York: Atlantic Books.

MacCannell, Dean (1976) *The Tourist: A New Theory of the Leisure Class*. New York: Schocken.

Marvin, Garry (2005) 'Guest Editor's Introduction: Seeing, Looking, Watching, Observing Nonhuman Animals'. *Society & Animals* 13.1: 1–12.

Melville, Herman (2003) *Moby-Dick; or, the whale*. London: Penguin Classics.

Morton, Timothy (2007) *Ecology without Nature: Rethinking Environmental Aesthetics*. Cambridge, MA: Harvard University Press.

Morton, Timothy (2010) *The Ecological Thought*. Cambridge, MA: Harvard University Press.

Morton, Timothy (2013) *Hyperobjects: Philosophy and Ecology after the End of the World*. Minneapolis: University of Minnesota Press.

Morton, Timothy (2016) *Dark Ecology: For a Logic of Future Coexistence*. New York: Columbia University Press.

Nyman, Jopi and Nora Schuurman (eds) (2016) *Affect, Space and Animals*. London: Routledge.

Orams, Mark B. (2000) 'Tourists Getting Close to Whales, Is It What Whale-watching Is All About?' *Tourism Management* 21: 561–569.

Peace, Adrian (2005) 'Loving Leviathan: The Discourse of Whale Watching in an Australian Eco-tourist Location'. In John Knight (ed.) *Animals in Person: Cultural Perspectives in Human-Animal Intimacies*. London: Berg, 191–210.

Pensky, Max (1993) *Melancholy Dialectics: Walter Benjamin and the Play of Mourning*. Amherst, MA: University of Massachusetts Press.

Pick, Anat (2011) *Creaturely Poetics: Animality and Vulnerability in Literature and Film*. New York: Columbia University Press.

Radden, Jennifer (ed.) (2000) *The Nature of Melancholy: From Aristotle to Kristeva*. Oxford: Oxford University Press.

Ross, Andrew (1991) *Strange Weather: Culture, Science, and Technology in the Age of Limits*. New York: Verso.

Rothenberg, David (2008) *Thousand Mile Song: Whale Music in a Sea of Sound*. New York: Basic Books.

Santner, Eric (2006) *On Creaturely Life: Rilke, Benjamin, Sebald*. Chicago: University of Chicago Press.

Shukin, Nicole (2009) *Animal Capital: Rendering Life in Biopolitical Times*. Minneapolis: University of Minnesota Press.

Silverman, Kaja (1996) *The Threshold of the Visible World*. New York: Routledge.

Stewart, Frank (ed.) (1995) *The Presence of Whales: Contemporary Writings on the Whale*. Anchorage: Alaska Northwest Books.

Van Dooren, Thom (2014) *Flight Ways: Life and Loss at the End of Extinction*. New York: Columbia University Press.

Weil, Kari (2010) 'A Report on the Animal Turn'. *Differences* 21.2: 1–23.

White, T.I. (2007) *In Defense of Dolphins: The New Moral Frontier*. Malden, MA: Blackwell.

Whitehead, Hal and Luke Rendell (2015) *The Cultural Lives of Whales and Dolphins*. Chicago: University of Chicago Press.

Woodward, Wendy (2001) *The Animal Gaze: Animal Subjectivities in Southern African Narratives*. Johannesburg: Witwatersrand University Press.

Postscript

In early February 2017, as this book was nearing completion, there were two separate mass strandings at Farewell Spit, a narrow sandspit at the northern end of Golden Bay on New Zealand's South Island. A total of 650 pilot whales were washed ashore, with over 300 of them dying despite a concerted rescue effort that succeeded in refloating some, while others were eventually able to conquer their exhaustion and swim back out. In the immediate aftermath of the strandings, beachgoers were warned to stay away from the dead whales, whose decomposing bodies, repositories for built-up gases, were in danger of exploding. Rescue teams were called upon for a second time to help with the clean up, distributing themselves among the whales and popping them like balloons to let the noxious gases out.

Sad spectacles of this kind are by no means uncommon occurrences. At the end of 2015, 337 sei whales died in a Chilean fjord in what was the largest ever recorded beaching of whales of this species (Evans 2017). On a different scale but still a record, twenty-nine sperm whales were found stranded in February 2016 alone on the shores of Britain, France, Germany, and the Netherlands (Evans 2017). Going back in time, there is substantial historical evidence of whale strandings worldwide, some of which were fearfully interpreted as apocalyptic portents, while others provided welcome opportunities for the exercise of royal authority: in medieval Denmark, says the American conservation biologist Joe Roman, 'all stranded whales were royal fish [since the king] owned the shore', while 'as far back as the fourteenth century, English and Scottish royalty had similar prerogatives over whales stranded or hunted along their coasts' (2006: 50).

It is still not known what causes whale strandings, although theories abound, several of them as distressing as the occurrences themselves. One almost certain factor is stress, however defined,[1] while another is the temporary disruption or sudden malfunctioning of whales' otherwise efficient 'biosonar' systems, complex echolocation devices designed to steer them, often for vast distances, through the ocean depths (see Chapter 1). A still more upsetting possibility is that whales – a nurturing kind – tend to protect one another, especially when one or more of them is in trouble, so the stranding of one whale can quite easily lead to the stranding of scores, even hundreds, at a time. There is an almost sacrificial dimension to this, reminding us perhaps of the time-honoured 'victim role' performed by animals, who die that we might live, and whose sanctified bodies come to operate as both material and symbolic substitutes for our own (Scholtmeijer 1993). There is thus something archaically satisfying, as well as viscerally upsetting, about the human witnessing of whale mass strandings – as if some kind of cleansing ritual is being enacted right in front of us; and as if, even though we may not be directly responsible for it, we are complicit nonetheless. This resonates to some extent with both Derrida's work on *sacrifice* – especially the much commented-on notion that 'the animal' must be sacrificed in order to create the conditions of possibility for 'the human' (Wolfe 2003) – and Agamben's work on *bare life*, which in Laura Hudson's useful gloss is 'neither precisely human nor animal [but emerges] at the moment that the concept of "the human" is produced through the separation of human from animal within the body of the human being' (Hudson 2008: n.p.; see also Agamben 2004). It should go without saying here that animal life (Agamben's *zōe*) is not *synonymous* with bare life, but can be *reduced* to the exploitative conditions that govern it: for example, via the powerful idea of the 'killable' animal (Lopez and Gillespie 2015).

These debates, which are central to human–animal studies as a philosophical field, inevitably raise the issue of animal rights – an issue that some might well see me as having studiously avoided in this book. I should probably make my own position clear here. Paola Cavalieri, one of the most vocal champions of animal rights today, dramatically announces that 'the history of human relationships with other cetaceans can [...] be seen as a genocidal history', suggesting that there is nothing in the term 'genocide' which 'logically prevents it from being applied to nonhuman beings' as well (2017: n.p.). Based on this appalling history, she further contends, it is possible to see whales and other cetaceans in terms of bare life, the extermination of which she interprets as nothing other than 'the actualization of [human or animal] victims' liability to be killed' (n.p.). This then provides the basis for her defence of cetacean rights, which, after the work of the American legal scholars Anthony D'Amato and Sudhir Chopra, she argues 'must be raised from the sphere of bare life governed by instrumental calculations to [a] favoured realm where rights and prerogatives apply' (n.p.). Rights need to be extended to cetaceans, she believes, insofar as they have a legitimate entitlement to personhood, which in turn suggests that new forms of global governance need to be found that take the regulation of human–cetacean relations away from the former hunting nations, which still dominate international bodies such as the International Whaling Commission (IWC).

It is difficult to disagree with Cavalieri's last point about the IWC, which is now widely seen as one of the most incompetent regulatory bodies in recent history (for a suitably derisive account of its many inconsistencies and misdemeanours, see Burnett 2012). The rest of her argument, however, is problematic on several different fronts. First and foremost, it is difficult to see how rights discourses can actually help cetaceans, based as these discourses generally are on the automatic, or at least uncritical, extension of *human* principles (e.g. the principle of individual freedom) into the *animal* domain. Second, as Cavalieri herself admits, the doctrine of human rights, even if it is extended to the domain of non-human animals, raises the question of *which* animals are considered to be qualified. Here, Cavalieri sensibly avoids the thorny debate around human/animal sentience, only to seek refuge in a hazy notion of intentionality, whereby the primary 'criterion for access to the enjoyment of human rights lies [...] in being an agent, that is, an intentional being that cares about its goals and wants to achieve them' (n.p.). This seems tantamount to according entitlement to the entire animal kingdom, not just to those animals that are deemed to share certain characteristics with human beings – which is itself a dubious criterion, for the reasons just given above. Third, Cavalieri's universalizing discourse of rights fails to recognize important cultural differences, for example, those surrounding aboriginal subsistence whaling, the grounds for which she summarily dismisses as an example of an atavistic appeal to 'cultural traditions' that stands in the way of a 'consistent application of undisputed moral doctrines' (n.p.) – as if it were enough to make a moral doctrine 'undisputed' simply by saying that it is. Finally, Cavalieri misreads Agamben in suggesting that animal life is, in effect, bare life: Agamben says no such thing, and, although it might be possible to question in turn some of Agamben's working distinctions, it is simply not right to use his arguments to make a case for 'ascribing fundamental protection to [all] defenceless beings' (n.p.), with 'defencelessness' again serving as an example of a human characteristic being uncritically projected onto the animal domain.

My point is hopefully made, but where does this leave us with cetaceans? The best preliminary move – for me at least – might be to recognize that whales are not like us at all, even if we co-inhabit the world with them (and with other animals). Here, I am partly following the Australian philosopher John Rundell, who argues that one of the founding principles for a new ethics of human–animal relations should be the recognition that 'we are largely unknown to one another' (2012: 18). Elaborating on this, he says that 'we, as humans, cannot enter the worlds of dogs, falcons or horses [still less the non-domesticated world of whales], even though they enter our own on our terms, for training, work [or] companionship. They have their own relationality, their own world and dynamics, which cannot be simply summed up in terms of our imagined [...] reconstruction of them through behavioural, ethological, or evolutionary sciences' (18–19). I would agree with this, as with the Levinasian conclusion that Rundell draws from it: that acknowledging the otherness of the Other can be the basis for acting responsibly towards the Other, not so much in terms of 'the fast, technically instituted time of progress, but [rather in] slow time – the time for different kinds of imaginings, reflection, contemplation, relationships and non-relationships with both human and non-human subjects' (Rundell 2012: 20; see also Levinas 1996).

I trust that this book, scattered though its speculations are, has opened up at least some of these disparate imaginings and reflections, which, in true melancholic fashion, are as likely to generate hope for a different kind of future as they are to feed despair in face of what appears to be an eternally self-perpetuating past. The point in any case is not to project our own fears and aspirations onto whales, although history has shown that we have been adept at doing so, but to try to understand the different worlds they inhabit, which are profoundly implicated in but by no means reducible to the world we all share. This 'multispecies' world – to use a term that has become common currency within animal studies – is based on the acknowledgement that there are many other forms of life besides the human, and that our lives are entangled with those of a multitude of different organisms whose existences are linked, whether directly or indirectly, to human social worlds (Helmreich and Kirksey 2010). However, the term 'entanglement' hints at a proximity that may not exist, a companionship that may turn out to be illusory. Whales, after all, are hardly companionate creatures, however much we might try to persuade ourselves that we share several important characteristics with them; nor are their worlds commensurable with our own, however much we might try to convince ourselves that these worlds are all effectively one and the same. Finally and most importantly, there is nothing inevitable about the modern capitalist world that has produced the figure of the melancholy whale any more than the historical slaughter of actual whales was an inevitable product of the needs of modern industrial capitalism. Perhaps – to return to the beginning of this book – it is time to consider conceptual alternatives to 'saving the whales', and to turn our attention instead to transforming the social and economic conditions that have historically governed our subordinating relationship with non-human animals, as well as with the human others that we either knowingly or involuntarily subordinate to ourselves.

Note

1 The most common dictionary definition of stress is 'exposure to unpleasant conditions with adverse effects' (quoted in Broom and Johnson 1993: 58), while more scientific approaches examine the physiological consequences of exposure to these conditions, generally known in the scientific literature as 'stressors', and the wide variety of so-called 'stress responses' that these equally variable conditions enact (58). However, scientists also readily admit that stress is, at best, an inconvenient term of convenience. That animals suffer is a fact, and that they suffer from something we might call 'stress' is also a given; but what *kinds* of stress they suffer from, and how that stress can be alleviated, is much harder to resolve. To date, most of the work on cetacean stress has focused on captive animals; see however Higham, Bejder, and Williams (2014) for some useful discussions of the possible causes and effects of stress on cetaceans in the wild.

Works Cited

Agamben, Giorgio (2004) *The Open: Man and Animal*. Palo Alto, CA: Stanford University Press.

Broom, D.M. and K.G. Johnson (eds) (1993) *Stress and Animal Welfare*. London: Chapman & Hall.

Burnett, D. Graham (2012) *The Sounding of the Whale: Science and Cetaceans in the Twentieth Century*. Chicago: University of Chicago Press.

Cavalieri, Paola (2017) 'From Bare Life to Nonhuman Others'. *Logos: A Journal of Modern Society & Culture,* http://www.logosjournal.com/cetaceans-bare-life-nonhuman-others.php, accessed 26 June 2017.

Evans, Peter (2017) 'What Causes Whale Mass Strandings?' *The Conversation*, 15 February, https://theconversation.com/what-causes-whale-mass-strandings-72985, accessed 26 June 2017.

Helmreich, Stefan and S. Eben Kirksey (2010) 'The Emergence of Multispecies Ethnography'. *Cultural Anthropology* 25.4: 545–574.

Higham, J.A., Lars Bejder, and Rob Williams (eds) (2014) *Whale-watching: Sustainable Tourism and Ecological Management*. Cambridge: Cambridge University Press.

Hudson, Laura (2008) 'The Political Animal: Species-Being and Bare Life'. *Mediations* 23.2. www.mediationsjournal.org/articles/the-political-animal, accessed 26 June 2017.

Levinas, Emmanuel (1996) *Basic Philosophical Writings*. Ed. A.T. Peperzak, S. Critchley, and R. Bernasconi. Bloomington, IN: Indiana University Press.

Lopez, Patricia and Kathryn A. Gillespie (eds) (2015) *Economies of Death: Economic Logics of Killable Life and Grievable Death*. London: Routledge.

Roman, Joe (2006) *Whales*. London: Reaktion Books.

Rundell, John (2012) 'Modernity, Humans and Animals – Tensions in the Field of the Technical-Industrial Imaginary'. *New Formations* 76.1: 5–20.

Scholtmeijer, Marian (1993) *Animal Victims in Modern Fiction: From Sanctity to Sacrifice*. Toronto: University of Toronto Press.

Wolfe, Cary (2003) *Animal Rites: American Culture, the Discourse of Species, and Posthumanist Theory*. Chicago: University of Chicago Press.

Index

CPSIA information can be obtained
at www.ICGtesting.com
Printed in the USA
LVHW081958190219
608041LV00012B/232/P

9 781350 010895